The Institute of Mathematics
and its Applications
Conference Series

The Institute of Mathematics and its Applications Conference Series

Previous volumes in this series were published by Academic Press to whom all enquiries should be addressed. Forthcoming volumes will be published by Oxford University Press throughout the world.

NEW SERIES

1. *Supercomputers and parallel computation* Edited by D. J. Paddon
2. *The mathematical basis of finite element methods*
   Edited by David F. Griffiths

The mathematical basis of finite element methods

with applications to partial differential equations

Based on lectures at an expository conference
organized by the Institute of Mathematics and its Applications
and held at Imperial College, University of London, 5–7 January 1983

Edited by

DAVID F. GRIFFITHS
University of Dundee

CLARENDON PRESS · OXFORD · 1984

Oxford University Press, Walton Street, Oxford OX2 6DP
London New York Toronto
Delhi Bombay Calcutta Madras Karachi
Kuala Lumpur Singapore Hong Kong Tokyo
Nairobi Dar es Salaam Cape Town
Melbourne Auckland
and associated companies in
Beirut Berlin Ibadan Mexico City Nicosia

Oxford is a trade mark of Oxford University Press

Published in the United States
by Oxford University Press, New York

British Library Cataloguing in Publication Data
The Mathematical basis of finite element methods:
with applications to partial differential
equations. – (The Institute of Mathematics and
its Applications conference series. New series; 2)
1. Finite element method
I. Griffiths, David F.   II. Series
515.3'53   TA347.F5
ISBN 0-19-853605-4

Printed in Great Britain by
The Thetford Press,
Thetford, Norfolk

PREFACE

The material in this volume is based on the lectures given at a short expository conference on The Mathematical Basis of Finite Element Methods with Applications to Partial Differential Equations held at Imperial College of Science and Technology, 5th - 7th January, 1983. The task of presenting the twelve talks at the conference fell on the shoulders of seven well-known contributors to the finite element literature, namely, Todd Dupont, A.R. Mitchell, K.W. Morton, P.A. Raviart, Dick Wait, J.R. Whiteman and O.C. Zienkiewicz. Their authoritative contributions, which led to such a successful meeting, are warmly acknowledged.

The chapters follow, by and large, the titles of the talks presented at the meeting. To avoid undue overlap with material that is currently present in many standard texts on finite element methods, it was decided that those chapters dealing with what has now come to be regarded as foundation material should provide only summaries of their particular areas, indicating where possible the background necessary for deeper study and references to the literature for further detail. The remaining chapters, dealing primarily with recent developments, are pursued more fully. It is hoped that this blend will make the mathematical aspects of finite element methods more readily accessible to newcomers to the field as well as providing a convenient source of reference for others.

Thanks are due to the staff of the Institute of Mathematics and its Applications for so ably handling the organization of the meeting and the typing of all the manuscripts.

D.F. Griffiths
Dundee

ACKNOWLEDGEMENTS

The Institute thanks the authors of the papers, the editor, Dr. D.F. Griffiths (University of Dundee) and also Mrs. J. Parsons, Miss D. Wright and Miss K. Jenkins for typing the papers.

CONTENTS

1.	Function spaces by R. Wait	1
2.	Conforming methods for self-adjoint elliptic problems by R. Wait	15
3.	A short survey of parabolic Galerkin methods by T. Dupont	27
4.	Nonconforming elements by D.F. Griffiths and A.R. Mitchell	41
5.	A-posteriori error estimation and adaptive mesh refinement in the finite element method by O.C. Zienkiewicz and A.W. Craig	71
6.	Finite element methods for non-self-adjoint elliptic and for hyperbolic problems: Optimal approximations and recovery techniques by K.W. Morton	91
7.	Mixed finite element methods by P.A. Raviart	123
8.	Curved elements by A.R. Mitchell	157
9.	Introduction to the treatment of singularities in elliptic boundary value problems using finite element techniques by J.R. Whiteman and K.T. Schleicher	169
	Index	185

CONTRIBUTORS

A.W. CRAIG; *Civil Engineering Department, University College of Swansea, Applied Science Building, Singleton Park, Swansea SA2 8PP.*

T. DUPONT; *Department of Mathematics, University of Chicago, Chicago, Illinois 60637, USA.*

D.F. GRIFFITHS; *Department of Mathematical Sciences, The University, Dundee DD1 4HN.*

A.R. MITCHELL; *Department of Mathematical Sciences, The University, Dundee DD1 4HN.*

K.W. MORTON; *Oxford University Computing Laboratory, Numerical Analysis Group, 8-11 Keble Road, Oxford OX1 3QD.*

P.A. RAVIART; *Analyse Numerique, Tour 55-56, 5 etage, Universite Pierre et Marie Curie, 4 Place Jussieu, 75230 Paris, France.*

K.T. SCHLEICHER; *Fachbereich Mathematik, Technische Hochschule Darmstadt, Schlossgartenstrasse 7, Darmstadt, West Germany.*

R. WAIT; *Department of Statistics and Computational Mathematics, University of Liverpool, Brownlow Hill, P.O. Box 147, Liverpool L69 3BX.*

J.R. WHITEMAN; *Institute of Computational Mathematics, Brunel University, Uxbridge, Middlesex UB8 3PH.*

O.C. ZIENKIEWICZ; *Civil Engineering Department, University College of Swansea, Applied Science Building, Singleton Park, Swansea SA2 8PP.*

FUNCTION SPACES

R. Wait

(Department of Statistics and Computational Mathematics, University of Liverpool)

1. INTRODUCTION

This chapter provides an introduction to some of the notation and results of functional analysis required in a mathematical study of the finite element solution of partial differential equations. The topics included are not intended to form a well balanced view of the subject, they are merely a selection of the more fundamental concepts on which an analysis of conforming finite elements for elliptic problems is based. The analysis of time dependent problems appears elsewhere (Chapter 3) and the chapters both on nonconforming methods (Chapter 4) and on methods for non self-adjoint problems (Chapter 6) introduce additional material as required.

The immediate goal is, therefore, to provide a mathematical framework for an error analysis with the hope that this will eventually lead to some error estimation.

It is assumed that there exists $u(\underline{x})$, the solution of a differential equation as yet unspecified, and a finite element approximation to $u(\underline{x})$ denoted by $U_h(\underline{x})$. An analysis of the error $u(\underline{x}) - U_h(\underline{x})$ involves properties of the solution $u(\underline{x})$, the approximation $U_h(\underline{x})$ and the method used to obtain $U_h(\underline{x})$. The method of approximation itself depends on the form of the differential equation. In order to estimate the error, it is necessary to be able to measure the size of the error function

$$(u - U_h)(\underline{x}) \equiv u(\underline{x}) - U_h(\underline{x}).$$

The underlying differential equation

$$Lu = f \quad \text{in } \Omega$$

involves a differential operator L that can be viewed as a mapping of one function u onto another function, Lu, that is not so smooth. The notation adopted in this chapter when defining such mappings is $L: u \mapsto Lu$. As this chapter is intended only as a very specialised and restricted introduction, it is assumed that readers not familiar with the basic concepts of metric spaces and normed linear spaces will first consult such texts as Aubin (1977), Sawyer (1978) or the early chapters of Griffel (1981), Milne (1980), Wouk (1979), Cryer (1983) or similar introductory material. We denote by $\mathbb{R}$ the space of real numbers, and by H a space of functions $u \equiv u(\underline{x})$, $v \equiv v(\underline{x})$, etc. the properties of which have yet to be specified. A linear functional T, maps H into $\mathbb{R}$ i.e. $T: H \to \mathbb{R}$. In finite element error analysis, functionals are usually formulated as definite integrals e.g.,

$$T(u) = \int_{\Omega} u(\underline{x})\,w(\underline{x})\,d\underline{x},$$

where $w(\underline{x})$ is a fixed weight-function. Given the definition of norms on two spaces H_1 and H_2, a mapping is said to be bounded if there exists $K > 0$ such that for all $u \in H_1$ $(T(u) \in H_2)$

$$\|T(u)\|_{H_2} \le K\|u\|_{H_1} \tag{1.1}$$

The smallest value of K that satisfies (1.1) for all $u \in H_1$, is identified as the norm of the mapping T. If the norm $\|\cdot\|$ is such that

$$\|T\| < 1$$

then, T is called a contraction mapping and if $H_1 = H_2$, the iteration

$$u^{(n+1)} = T\left(u^{(n)}\right) \qquad n = 1,2,..$$

converges to the solution of the equation

$$u = T(u).$$

Equivalent definitions of norms are:

(i)

$$\|T\| = \sup_{\|u\|\le 1} \{\|T(u)\|\}$$

(ii)

$$\|T\| = \inf_{u \in H} \{K > 0: \|T(u)\| \le K\|u\|\}$$

(iii)

$$\|T\| = \sup_{\|u\|\neq 0} \left\{\frac{\|T(u)\|}{\|u\|}\right\}. \tag{1.2}$$

Boundedness is a necessary and sufficient condition for continuity of the mapping and we shall use the two expressions interchangeably. A continuous mapping T is such that, if the sequence $\{u_n\}$ converges to u in H_1, then the sequence $\{T(u_n)\}$ converges to $T(u)$ in H_2. Most of the elementary results of this type that are required in finite element analysis are collected together in Mitchell and Wait (1984).

2. HILBERT SPACES

Before proceeding to any finite element analysis, it is necessary to identify an important set of vector spaces known as Hilbert spaces. In the next section attention will be restricted still further to a subset of spaces known as Sobolev spaces, wherein lie the solutions of elliptic differential equations.

A Hilbert space H, in addition to the properties common to all normed linear spaces, possesses an inner product $(,):H \to \mathbb{R}$. Thus, for any pair $u,v \in H$, the real number (u,v) is uniquely defined such that

(i) $(u,v) = (v,u)$

(ii) $(\alpha u_1 + \beta u_2, v) = \alpha(u_1,v) + \beta(u_2,v) \qquad \alpha,\beta \in \mathbb{R}.$

The norm is then defined as

$$\|u\| = (u,u)^{\frac{1}{2}}.$$

An inner product is an example of a bilinear form, that is, it is linear in each of the arguments u and v, and it delivers a real number as a result. In a finite element context, the inner product of a typical Hilbert space H might be

$$(u,v) = \int_{-1}^{1} \left\{ u(x)v(x) + \frac{du}{dx}\frac{dv}{dx} \right\} dx. \tag{2.1}$$

In addition a Hilbert space is complete, i.e. it must contain the limit points of all bounded sequences. Thus if it is possible to construct a sequence of smooth functions $u_n(x) \in H$, where

$$\|u_n\| \le c,$$

with $\|\cdot\|$ defined in terms of the inner product (2.1), for which

$$\lim_{n\to\infty} u_n(x) = u(x)$$

then $u(x) \in H$. Hence with the inner product (2.1), piecewise smooth functions (Fig. 1a) are in H, but the Heaviside function (Fig. 1b) is not. Readers unsure of the notions of convergence and weak convergence should consult the references.

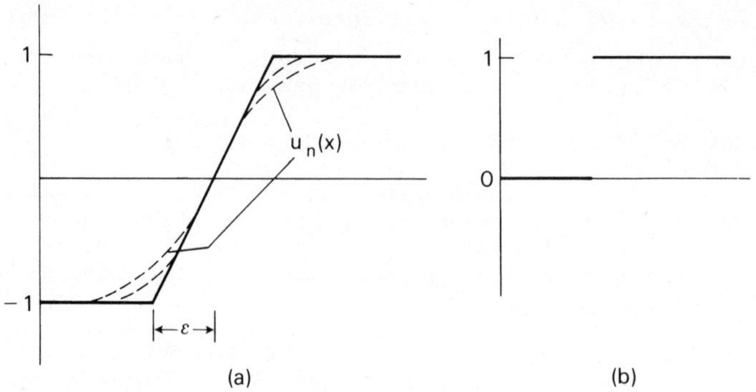

Fig. 1 (a) Piecewise linear u_ε (b) Heaviside function

The space of bounded linear functionals on H, known as the dual space of H and denoted by H*, has many of the properties of the space H itself. It is possible to identify each element of the dual space H* with a unique element of the primal space H.

Reisz Representation theorem

For any Hilbert space H and for each $T \in H^*$, there exists a unique $u \in H$, such that

$$T(v) = (u,v) \qquad \text{for all } v \in H.$$

The mapping $J:u \mapsto T$ from H to H* is known as the Reisz map and $u \in H$ is known as the Reisz representor of the functional T.

The proof is given, inter alia, by Showalter (1977) and Griffel (1981).

Every inner product must satisfy the Cauchy-Schwartz inequality

$$(u,v) \leq \|u\| \; \|v\|$$

and so from (1.2),

$$\|T\| \leq \|u\| .$$

But, with v=u, it follows that $\|T\| \geq \|u\|$ and so

$$\|T\| = \|u\| .$$

It follows from a second application of (1.2) that $\|J\| = 1$ and J is an isometry between H and H*, i.e. it preserves the structure.

Inner products are not the only bilinear forms that are of importance in finite element analysis. A general bounded bilinear form $A(,): H \times H \to \mathbb{R}$ is such that there exists $K > 0$ for which

$$|A(u,v)| \leq K\|u\|\|v\|, \tag{2.2}$$

$$A(u,) \in H^* \qquad \text{for all } u \in H$$

and

$$A(,v) \in H^* \qquad \text{for all } v \in H.$$

An additional property that arises naturally and proves useful in the analysis is that for some $\alpha > 0$

$$A(u,u) \geq \alpha\|u\|^2 \qquad \text{for all } u \in H, \tag{2.3}$$

this is known as coercivity or H-ellipticity or positive definiteness of A. If the bilinear form A(,) is also symmetric, i.e.

$$A(u,v) = A(v,u),$$

as well as being coercive, then it is equivalent to an inner product. It is then possible to define an energy space comprising the components of the space H, equipped with the energy inner product

$$(u,v)_A \equiv A(u,v)$$

and it follows that there is the corresponding energy norm

$$\|u\|_A = [A(u,u)]^{\frac{1}{2}},$$

for which the Cauchy-Schwartz inequality holds in the form

$$|(u,v)_A| \leq \|u\|_A\|v\|_A.$$

In later chapters, finite element approximations are applied to differential equations in the weak form where they are formulated as

$$A(u,v) = F(v) \qquad \text{for all } v \in H. \tag{2.4}$$

The nature of the bilinear form A(,), the functional F and the Hilbert space H are determined by the differential equation and the boundary conditions. A basic result is clearly whether or not a solution exists when the problem is formulated as (2.4).

Lax-Milgram Lemma

If A(,) is a bounded coercive bilinear form on a Hilbert space H, and $F \in H^*$ then there exists a unique $u \in H$ such that (2.4) holds.

A proof can be found in Ciarlet (1978) or Showalter (1977) and it uses the Reisz representation theorem to construct a representor $A_u \in H$ of $A(u,) \in H^*$, then defines a mapping $T:H \to H$ such that $T(u) = A_u$. Similarly, $f \in H$ is the Reisz representor of $F \in H^*$. It is then shown that the mapping

$$v \mapsto v - \rho(T(v) - f)$$

is a contraction map for some ρ and hence the equation

$$T(v) - f = 0 \tag{2.5}$$

has a unique solution. The proof is then completed by showing that (2.5) and (2.4) are equivalent.

In a finite element context, when (2.4) is the weak (or integrated) form of the differential equation, (2.5) is the classical form. A conforming finite element approximation U_h to (2.4), satisfies a similar equation, viz

$$A(U_h,V_h) = F(V_h) \qquad \text{for all } V_h \in K \tag{2.6}$$

where V_h is now restricted to a subspace K of the test space H and U_h is the subspace approximation to u.

Using the weak forms such as (2.4), it is possible to devise error bounds in terms of the energy norm and then to translate these into bounds in terms of norms derived from inner products such as (2.1), i.e. involving derivatives.

Cea's Lemma (Ciarlet (1978))

If the bilinear form A(,) is H-elliptic and bounded in terms of the norm $\|\cdot\|$, and if $u \in H$ and $U_h \in K \subset H$ are defined as the solutions of (2.4) and (2.6) respectively, then U_h is the best approximation to u in terms of the energy norm $\|\cdot\|_A$, i.e.

$$\|u-U_h\|_A = \inf_{V_h \in K} \|u - V_h\|_A$$

and is an optimal approximation in terms of $\|\cdot\|$, i.e.

$$\|u-U_h\| \le \frac{K}{\alpha} \inf_{V_h \in K} \|u - V_h\|,$$

where K and α are the constants defined in (2.2) and (2.3) respectively. The proof follows immediately from the definitions once it is observed that (2.4) and (2.6) imply

$$A(u - U_h, V_h) = 0 \quad \text{for all } V_h \in K,$$

thus

$$A(u - U_h,\ u - U_h) = A(u - U_h,\ u - V_h)$$

$$+ A(u - U_h,\ V_h - U_h),$$

where the second term on the right is zero. The desired results then follow by applying ellipticity on the left and continuity on the right.

If, alternatively, an error bound is required in terms of the L_2-inner product (say)

$$(u,v) = \int_{-1}^{1} u(x)v(x)dx, \tag{2.7}$$

it is necessary to make use of the so-called Aubin-Nitsche lemma. If we consider functions with the inner product (2.7), then the corresponding Hilbert space H_0 must include functions such as the Heaviside function of Fig. 1(b), that were excluded from H_1, the Hilbert space with inner product (2.1); thus $H_1 \subset H_0$. It can be seen from Fig. 1 that it is possible to get arbitrarily close to the Heaviside function using piecewise linear functions that are in H_1, but the sequence is unbounded since $\|U_\varepsilon\|_{H_1} \to \infty$ as $\varepsilon \to 0$. Thus we say that H_1 is embedded in H_0 and that it is a dense embedding (written $\bar{H}_1 = H_0$).

Aubin - Nitsche Lemma

If $H_1 \subset H_0$ and $\bar{H}_1 = H_0$ with norms $\|\cdot\|_1$ and $\|\cdot\|_0$ respectively and if in addition, $u \in H_1$ satisfies (2.4) while $U_h \in K_1 \subset H_1$ satisfies (2.6) with $A(,)$ a bounded bilinear form on H_1, it then follows that

$$\|u - U_h\|_0 \leq K\|u - U_h\|_1 \left\{ \sup_{G \in H_0^*} \left[\frac{1}{\|G\|} \inf_{V_h \in K_1} \|w - V_h\|_1 \right] \right\} \tag{2.8}$$

where for each $G \in H_0^*$, $w \in H_1$ satisfies

$$A(v,w) = G(v) \quad \text{for all } v \in H_1 \tag{2.9}$$

Proof

The proof is very short and can also be found in both Ciarlet (1978) and Showalter (1977). It is included here as similar arguments are used in a number of different areas of finite element analysis.

For a space such as $H_0 (=L_2)$, a so-called pivot space, it is possible to invert the definition (1.2) for linear functionals to give a dual definition

$$\|u\|_H = \sup_{T \in H^*} \frac{|T(u)|}{\|T\|_{H^*}}$$

and hence in this case

$$\|u - U_h\|_0 = \sup_{G \in H_0^*} \frac{|G(u - U_h)|}{\|G\|} . \tag{2.10}$$

Since

$$u - U_h \in H_1$$

it follows from (2.9) that

$$G(u - U_h) = A(u - U_h, w)$$

and as before, since $V_h \in K_1 \subset H_1$, it follows, by subtracting (2.6) from (2.4), that

$$A(u - U_h, V_h) = 0.$$

Thus

$$G(u - U_h) = A(u - U_h, w - V_h)$$

and so

$$|G(u - U_h)| \leq K\|u - U_h\|_1 \|w - V_h\|_1 .$$

As V_h is arbitrary, this can be replaced by

$$|G(u - U_h)| \leq K\|u - U_h\|_1 \inf_{V_h \in K_1} \|w - V_h\|_1 ,$$

then, combining this with (2.10) leads to the desired result viz (2.8).

3. SOBOLEV SPACES AND DISTRIBUTIONS

If we denote by Ω the bounded domain over which the differential equation is specified, then in order to apply the results of the preceding section, it is necessary to define spaces of functions defined on the domain Ω. Let $C^\infty(\Omega)$ be the space of smooth functions with smooth derivatives of all orders defined on Ω, then $C_0^\infty(\Omega) \subset C^\infty(\Omega)$ contains those functions that also vanish on Γ, the boundary of Ω. Such functions are said to have compact support; an alternative notation for $C_0^\infty(\Omega)$ is $\mathcal{D}$. A distribution (or generalised function) is a continuous linear functional on $\mathcal{D}$, thus $\mathcal{D}^*$ is the space of distributions (see Ciarlet (1978), Oden and Reddy (1976) or Showalter (1977)). The space L_2 is the completion of $\mathcal{D}$ in terms of the inner product

$$(u,v) = \int_\Omega u\ v\ d\underline{x}$$

and hence of the norm

$$\|u\| = \left\{ \int_\Omega u^2 d\underline{x} \right\}^{\frac{1}{2}}.$$

As stated in the preceding section, L_2 includes discontinuous functions such as the Heaviside function. If we identify $L_2^* = L_2$ using Reisz representors, then

$$L_2(\Omega) = C_0^\infty(\Omega) \quad + \text{distributions.}$$

If $\{\phi_n\} \subset \mathcal{D}$ such that (ϕ_n,ψ) converges for all $\psi \in \mathcal{D}$ i.e. $\{\phi_n\}$ is weakly convergent, then we can define the distribution $\Phi:\psi \to \lim_{n\to\infty} (\phi_n,\psi) = (\Phi,\psi)$ using the Reisz representation theorem. As we have now defined distributions in terms of smooth functions, it is possible to extend the definition to provide derivatives of distributions. Thus, in one dimension,

$$(\phi',\psi) = -(\phi,\psi') \qquad \phi,\psi \in \mathcal{D}$$

and it is possible to define Φ' such that

$$(\Phi',\psi) = -\lim_{n\to\infty} (\phi_n,\psi')$$

$$= -(\Phi,\psi') \qquad \text{for all } \psi \in \mathcal{D}.$$

In particular, the derivative of the Heaviside function is the Dirac delta function (see Oden and Reddy, 1976, or Treves, 1980, for a more complete account of distributions).

Sobolev spaces are Hilbert spaces of functions, equipped with inner products such as (2.1). The norms and inner products are best described in terms of the standard multi-index notation (Mitchell and Wait (1984) or Oden and Reddy (1977)) for derivatives of functions of many variables. Thus if $\Omega \subset \mathbb{R}^N$ we define

$$D_i \equiv \frac{\partial}{\partial x_i} \quad (i = 1,\dots,N)$$

and

$$D^k \equiv D_1^{k_1} \dots D_N^{k_N} \quad (k \equiv k_1,\dots,k_N : k_1 \geq 0).$$

if

$$|k| = k_1 + \dots + k_N,$$

then

$$|u|_{j,p,\Omega} \equiv \left\{ \sum_{|k|=j} \iint_\Omega |D^k u|^p \right\}^{1/p}$$

is a semi-norm and

$$\|u\|_{m,p,\Omega} \equiv \left\{ \sum_{j=0}^{m} |u|^p_{j,p,\Omega} \right\}^{1/p}$$

is the norm for the Sobolev space $W^{m,p}(\Omega)$ which can be defined as the completion of $C^\infty(\Omega)$ with respect to the norm $\|\cdot\|_{m,p,\Omega}$. It is usual to employ the alternative notation H^m for the spaces $W^{m,2}$. It is possible to define Sobolev spaces with fractional indices (see Grisvard (1975) or Adams (1975)) using the theory of intermediate spaces to arrive at a definition of the so-called Besov spaces. The spaces $H_0^m(\Omega)$ are the completion of $\mathcal{D}$ in terms of the norms $\|\cdot\|_{m,\Omega} \equiv \|\cdot\|_{m,2,\Omega}$ and $H^{-m}(\Omega) \equiv (H_0^m(\Omega))^*$ (see Oden and Reddy (1976) or Ciarlet (1978)). With intermediate spaces it is possible to identify the reduction in continuity inherent in the Trace operators (Showalter (1977)) that map a function defined throughout the region Ω to a function defined only on the boundary Γ. The boundary spaces $W^{m,p}(\Gamma)$ are defined in terms of tangential derivatives, integrated over the manifold $\Gamma \subset \mathbb{R}^{N-1}$. Continuity on Γ is less restrictive than continuity throughout Ω and it can be shown (Aubin (1979)), that

$$K_1\|v\|_{m,p,\Omega} \le \|u\|_{m-1/p,p,\Gamma} \le K_2\|v\|_{m,p,\Omega}$$

where $u \equiv v|_{\Gamma}$. Hence an equivalent definition of the norm on the boundary space $H^{m-\frac{1}{2}}(\Gamma)$ in terms of $H^m(\Omega)$, is

$$\|u\|_{m-\frac{1}{2},\Gamma} = \inf\left\{\|v\|_{m,\Omega} : v|_{\Gamma} = u\right\}.$$

If the boundary Γ can be written (locally) as a manifold

$$x_N = f_N(x_1,\ldots,x_{N-1}),$$

and if f_N is locally Lipschitz continuous, we say that Γ is locally Lipschitz continuous (Nečas (1967). The trace operator $\gamma_0 : C^1(\bar{\Omega}) \to C^0(\Gamma)$ is such that

$$(\gamma_0 u)(x_1,\ldots,x_{N-1}) = u(x_1,\ldots,x_{N-1}, f_N).$$

An alternative formulation of the geometrical properties required of the boundary is the so-called cone condition (Adams (1976), Oden and Reddy (1976)). When the boundary has sufficient smoothness to satisfy these conditions (however formulated) it is possible to replace estimates in terms of Sobolev norms involving integrals, i.e. mean values in some sense, by maximum norm estimates via the Sobolev inequality

$$\|u\|_{k,\infty,\Omega} \le C\|u\|_{m,\Omega} \qquad (m > k + \tfrac{N}{2})$$

when it is possible to apply the Sobolev Embedding theorem (Adams (1976), Oden and Reddy (1976)).

The basic result concerning the estimation of errors in polynomial approximation, in terms of Sobolev norms, was provided by Bramble and Hilbert (1970). It forms the basis for nearly all finite element error bounds in terms of Sobolev norms or semi-norms. Earlier results had an error analysis constructed in terms of maximum norm bounds and it is clear from the Sobolev inequality that an estimate in terms of the norm $\|\cdot\|_{m,\Omega}$ is a more powerful and less restrictive than one in terms of the norm $\|\cdot\|_{m,\infty,\Omega}$ that occurs naturally using simple bounds on the remainder in Taylor series.

Bramble-Hilbert lemma

Let $F \in W^{k+1,p}(\Omega)^*$ be such that

$$F(u) = 0 \qquad \text{for all } u \in P_k$$

i.e. F annihilates all polynomials of degree k or less. If Ω satisfies the strong cone condition then

$$|F(u)| \leq C\|F\| \, |u|_{k+1,p,\Omega}.$$

In a finite element context, this inequality is applied on each element, so that the geometric conditions on Ω are always satisfied.

The proof (see Bramble and Hilbert (1970), Ciarlet (1978), Mitchell and Wait (1984), Durán (1983) or Dupont and Scott (1980)) uses the observation that

$$F(u) \doteq F(u+q) \qquad \forall q \in P_k.$$

Then

$$|F(u)| \leq \|F\| \|u+q\|$$

and using quotient spaces $W^{k+1,p}/P_k$, it can be shown that

$$\inf_q \|u+q\| \leq C|u|_{k+1,p,\Omega}.$$

4. CONCLUSION

This chapter has attempted to provide a brief survey of the mathematical topics that are most frequently introduced during the analysis of finite element methods for linear problems. No attempt has been made to cover nonlinear problems, examples of which can be found in Varga (1971); the basic theory is introduced in Sawyer (1978) or Wouk (1979). The treatment here, particularly that of distributions and trace operators, has been extremely superficial, a more detailed account can be found in the references.

REFERENCES

Adams, R.A. (1975) Sobelev spaces, New York; Academic Press.

Aubin, J.-P. (1977) Applied Abstract Analysis, New York; Wiley.

Aubin, J.-P. (1979) Applied Functional Analysis, New York; Wiley.

Bramble, J.H. and Hilbert, S.R. (1970) Estimation of linear functionals with applications to Fourier transforms and spline interpolation. *SIAM J. Num. Analysis,* **7**, 112-124.

Ciarlet, P.G. (1978) The Finite Element Method for Elliptic Problems, Amsterdam; North-Holland.

Cryer, C.W. (1983) Numerical Functional Analysis, Oxford; OUP.

Dupont, T. and Scott, R. (1980) Polynomial approximation of functions in Sobolev spaces. *Math. Comp.,* **34**, 441-463.

Durán, R.G. (1983) On polynomial approximation in Sobolev spaces. *SIAM J. Num. Analysis*, **20**, 985-988.

Griffel, D.H. (1981) Applied Functional Analysis, Chichester; Ellis Horwood.

Grisvard, P. (1975) Behavior of the solution of an elliptic boundary value problem in a polynomial or polyhedral domain. In "Numerical solution of partial differential equations - III, SYNSPADE 75", (B. Hubbard, Ed.), New York; Academic Press.

Milne, R.D. (1980) Applied Functional Analysis, London; Pitman.

Mitchell, A.R. and Wait, R. (1984) The Finite Element Method in Partial Differential Equations, Chichester; Wiley.

Nečas, J. (1967) Les Methodes Directes en Theorie des Equations Elliptiques, Paris; Masson.

Oden, J.T. and Reddy, J.M. (1976) An Introduction to the Mathematical Theory of Finite Elements, New York; Wiley.

Sawyer, W.W. (1978) A First Look at Numerical Functional Analysis, Oxford; OUP.

Showalter, R.E. (1977) Hilbert Space Methods for Partial Differential Equations, London; Pitman.

Treves, F. (1980) Introduction to Pseudodifferential and Fourier Integral Operators, New York; Plenum.

Varga, R.S. (1971) Functional Analysis and Approximation Theory in Numerical Analysis, Philadelphia; SIAM.

Wouk, A. (1979) A Course in Applied Functional Analysis, New York; Wiley.

CONFORMING METHODS FOR SELF-ADJOINT ELLIPTIC PROBLEMS

R. Wait

(University of Liverpool)

1. INTRODUCTION

This chapter completes the basic analysis of chapter 1. Convergence estimates are derived in terms of Sobolev norms, L_2 norms and L_∞ norms. The conditions imposed for these results to apply mean that, in general, the solution has to be smooth, the integrals are performed analytically over the exact domain and that the essential boundary conditions are satisfied exactly. The method of approximation is assumed to be the standard Galerkin formulation, alternative formulations will be mentioned in later chapters. Where the results can be extended to encompass boundary approximations and numerical integration, the references are included. All elements are assumed to conform to the restrictions imposed by the theory of chapter 1.

2. SELF-ADJOINT BOUNDARY VALUE PROBLEMS

The classical form of a differential equation can be written as

$$\mathcal{L}u = f \qquad \text{in } \Omega, \tag{2.1}$$

if this is of order m in N variables, the operator $\mathcal{L}$ can be expressed as

$$\mathcal{L} = \sum_{|k|\leq m} a_k(\underline{x})D^k ,$$

where D^k, for a multi-index k is defined in chapter 1. The ellipticity or otherwise of the operator $\mathcal{L}$ can be determined by examining the characteristics of the principal part

$$\mathcal{L}_p = \sum_{|k|=m} a_k(\underline{x})D^k,$$

see for example Stakgold (1979). A well posed boundary value problem consists of an elliptic equation (2.1), together with additional conditions imposed on the solution u at the boundary Γ of the domain Ω. The boundary conditions considered in later chapters vary slightly, but a typical set for a second order problem would be u specified on part of Γ with $\partial u/\partial n$ (the outward normal derivative) specified on the remainder. The solution can be defined in terms of the free-space Green's function $g(\underline{x},\underline{\xi})$ such that

$$\mathcal{L}\, g(\underline{x},\underline{\xi}) = \delta(\underline{x}-\underline{\xi})$$

where δ is the Dirac delta function. In $\mathbb{R}^2$ when L is the Laplace operator,

$$g(\underline{x},\underline{\xi}) = \ell n|\underline{x}-\underline{\xi}|$$

and this leads to a factor $\ell n(h)$ that appears in some of the L_∞ convergence estimates of section 5. Further details of this classical style of analysis can be found in Stakgold (1979), but finite element methods are based on the alternative weak form. Following the notation of chapter 1, this can be written as

$$A(u,v) = F(v) \qquad \text{for all } v \in H. \tag{2.2}$$

where in this case, the functional F is defined in terms of the L_2 inner product

$$(f,v) \equiv \int_\Omega f(\underline{x})\ v(\underline{x})d\Omega.$$

The bilinear form A(,) can be derived from the operator L using Green's theorem such that the difference between A(u,v) and (Lu,v) can be expressed solely in terms of boundary integrals, see Mitchell and Wait (1984) or Showalter (1977). The boundary value problem based on (2.1) and the weak form (2.2) are only equivalent if the trial space (of admissible functions u) and the test space (of admissible functions v) are defined correctly.

For simplicity, consider the Poisson problem

$$-\nabla^2 u = f \qquad \text{in } \Omega, \tag{2.3}$$

together with

$$u = g \qquad \text{on } \Gamma_1 \tag{2.4}$$

and

$$\frac{\partial u}{\partial n} = h \qquad \text{on } \Gamma_2 \tag{2.5}$$

where $\Gamma = \Gamma_1 \cup \Gamma_2$ and $\Gamma_1 \cap \Gamma_2 = \emptyset$. The Dirichlet condition (2.4), is an example of a forced or essential boundary condition that is imposed by specified subsidiary conditions on the test and trial spaces. In this example, the trial space is

$$H_1 = \{u: u \in H^1(\Omega) \text{ and } u = g \text{ on } \Gamma_1\},$$

if $g \not\equiv 0$, this is a linear manifold, not a linear space as defined in chapter 1. The test space is

$$H_0 = \{v: v \in H^1(\Omega) \text{ and } v = 0 \text{ on } \Gamma_1\}.$$

The Neumann condition (2.5), is an example of a natural boundary condition imposed via the functional F in (2.2). Thus the equivalent weak formulation requires

$$A(u,v) = \int_\Omega \left\{ \frac{\partial u}{\partial x}\frac{\partial v}{\partial x} + \frac{\partial u}{\partial y}\frac{\partial v}{\partial y} \right\} d\Omega \tag{2.6}$$

and

$$F(v) = (f,v) - \int_{\Gamma_2} hv \, d\Gamma. \tag{2.7}$$

Further details of the identification of natural and essential boundary conditions can be found in Mitchell and Wait (1984).

Problems for which the bilinear form is symmetric, i.e.

$$A(u,v) = A(v,u)$$

correspond to self-adjoint problems when written in classical form. If, in addition, $H_1 = H_0$, a boundary value problem can be formulated as a variational principle such that the solution u provides a stationary value for the quadratic functional

$$J(v) = A(v,v) - 2F(v).$$

If the bilinear form A(,) is coercive then the stationary value is an extremum.

3. CONFORMING GALERKIN APPROXIMATIONS

A Galerkin approximation is based on the weak form (2.2) by defining an approximate solution U_h as a finite sum of specified basis functions ϕ_i, i.e.

$$U_h = \sum_i U_i \phi_i \tag{3.1}$$

and the test functions are the subset $S_0 \subset H_0$ of combinations of basis functions that satisfy the homogeneous essential boundary conditions. Thus some of the coefficients U_i are specified by the boundary data, as a conforming approximation must satisfy $U_h \in H_1$, whilst the remainder are specified by the Galerkin equations

$$A(U_h,\phi_i) = F(\phi_i) \quad \text{for all } \phi_i \in S_0 \subset H_0.$$

(c.f. chapter 1 (2.6)). If the bilinear form A(,) is coercive, as in (2.6), then it is possible to apply Cea's lemma so that the Galerkin approximation to (2.3) - (2.5) is best in terms of the energy norm

$$\|\cdot\|_A = |\cdot|_{1,\Omega}$$

and optimal in terms of the Sobolev norm $\|\cdot\|_{1,\Omega}$, i.e.

$$\|u - U_h\|_{1,\Omega} \leq \frac{K}{\alpha} \inf_{V_h \in S_1} \|u - V_h\| \tag{3.2}$$

where $S_1 \subset H_1$ is the space of all approximating functions satisfying the inhomogeneous essential boundary conditions. If approximations that do not satisfy the boundary conditions exactly are to be included then it is necessary to use a perturbation analysis, see for example Mitchell and Wait (1984), or to use the interior estimates of section 5.

If a convergence estimate in terms of the L_2 norm is required, then it is necessary to use the Aubin-Nitsche lemma. We define the normalised error (Schultz, 1971)

$$e = \frac{u - U_h}{\|u - U_h\|_{0,\Omega}},$$

then define $G \in L_2^*(= L_2)$ as

$$G(v) = (e,v).$$

If this is used in the Aubin-Nitsche lemma ((2.8) in chapter 1), then, as

$$\|G\|_0 = \|e\|_0 = 1,$$

we have the bound

$$\|u - U_h\|_{0,\Omega} \leq K\|u - U_h\|_{1,\Omega} \|w - W_h\|_{1,\Omega}, \tag{3.3}$$

where $w \in H_0$ satisfies

$$A(v,w) = G(v) \qquad \text{for all } v \in H_0 \tag{3.4}$$

and $W_h \in S_0$ satisfies

$$A(V_h,W_h) = G(V_h) \qquad \text{for all } V_h \in S_0. \tag{3.5}$$

In the next section it is shown how the bound (3.2) is modified by the term $\|w - W_h\|_{1,\Omega}$ from (3.5) to provide an optimal L_2 - estimate from (3.3). All these results assume that the integrations are performed analytically; if numerical quadrature is used then a scheme that preserves coercivity has to be adopted (see Mitchell and Wait, 1984 or Ciarlet, 1978).

4. PIECEWISE POLYNOMIAL FINITE ELEMENTS

Conforming approximations require the partition of Ω into non-overlapping elements with no boundary approximation. This is a highly restrictive condition on non-polygonal domains as curved elements do not, in general, recover arbitrarily curved boundaries. If the union of the elements is a region $\Omega_h \neq \Omega$, then again a perturbation analysis is necessary to complete the convergence proofs (see Mitchell and Wait, 1984) unless the interior estimates of section 5 are acceptable.

Within each element, the basis functions (or trial/test functions) are defined locally in terms of shape functions that are specified by nodal parameters. The nodes are located in a way that guarantees the necessary degree of global continuity. Relaxing global continuity generates non-conforming elements, i.e. $S_0 \not\subset H_0$ and $S_1 \not\subset H_1$, for which the preceding estimates are no longer valid. A discussion of the theory and practice relating to non-conforming elements is given in chapter 4. Extensive catalogues of many of the different forms of elements can be found in Mitchell and Wait (1984) or Ciarlet (1978). The simpler types include triangular elements with linear shape functions defined in terms of the values at nodes located at the vertices, quadratic shape functions with additional nodes located at the mid-sides or rectangular elements with bilinear shape functions. Elements with curved sides can be constructed as the isoparametric extensions of polynomial elements. That is, polynomial shape functions ϕ_i^ε are defined on a reference element ε with nodes $\underline{\xi}_i$ and then a typical physical element E, defined in terms of nodes $\underline{x}_i^E$, is mapped onto ε using the transformation

$$T: \underline{\xi} \to \underline{x} = \sum_i \phi_i^\varepsilon \underline{x}_i^E.$$

The approximate solution is then defined locally in terms of the same parameters as

$$U^E = \sum_i \phi_i^\varepsilon U_i .$$

Convergence estimates are derived by combining the inequalities (3.2) and (3.3) with the Bramble-Hilbert lemma of chapter 1. If V_h in (3.2) is taken to be the 'finite element' interpolant of the solution, i.e. using the nodal values $u(\underline{x}_i^E)$, we can apply the Bramble-Hilbert lemma to the interpolant error in the $\|\cdot\|_{1,\Omega}$ norm. Piecewise polynomial elements recover low order polynomial solutions exactly, but isoparametric elements recover only linear solutions on Ω. Higher order polynomials in terms of the reference element only are recovered locally.

We require estimates of the form

$$\|u - U_h\|_{1,\Omega} = O(h^m)$$

where h is the diameter of a typical element E. To achieve this, we apply the Bramble-Hilbert lemma on the reference element, the transformation $T:\varepsilon \to E$ provides the scaling by the appropriate powers of h and we then sum over all elements to obtain the desired result. In order to complete the proof it is necessary to assume that such scaling factors do exist. Thus if Cea's lemma is valid in terms of the norm $\|\cdot\|_{m,\Omega}$, <u>i.e.</u> for a differential equation of degree 2m, then

$$\|u - V_h\|_{m,\varepsilon} \geq Ch^{m-1} \|u - V_h\|_{m,E} \tag{4.1}$$

this is often replaced by the equivalent assumption

$$\sum_{\ell \leq m} h^{\ell} |u - V_h|_{\ell,\varepsilon} \geq Ch^{m-1} \|u - V_h\|_{m,E} . \tag{4.2}$$

If the shape functions recover polynomials of degree k on ε then the Bramble-Hilbert lemma is expressed in terms of $|u|_{k+1,\varepsilon}$ and so we require an inequality of the form

$$|u|_{k+1,\varepsilon} \leq Ch^{k} \|u\|_{k+1,E} \tag{4.3}$$

to complete the proof. The right hand side of (4.3) can be replaced by $Ch^k|u|_{k+1,E}$ if the transformation $T:\varepsilon \to E$ is linear. The proof for linear transformations can be found in Ciarlet and Raviart (1972), for nonlinear isoparametric transformations the proof is more complex and assumes that the elements curvatures are small (see Mitchell and Wait, 1984). Cea's lemma states that

$$\|u - U_h\|_{m,\Omega} \leq C \inf_{V_h} \|u - V_h\|_{m,\Omega}$$

in particular if V_h interpolates u then

$$\|u - U_h\|_{m,\Omega} \leq C \|u - V_h\|_{m,\Omega}$$

where

$$\|\cdot\|_{m,\Omega} = \left\{ \sum_{E} \|\cdot\|^2_{m,E} \right\}^{\frac{1}{2}}.$$

From (4.1)

$$\|u - V_h\|_{m,E} \leq Ch^{1-m} \|u - V_h\|_{m,\varepsilon} ,$$

applying the Bramble-Hilbert lemma yields

$$\|u - V_h\|_{m,\varepsilon} \leq C\, |u|_{k+1,\varepsilon}\ ,$$

applying (4.3) leads to

$$\|u - V_h\|_{m,E} \leq Ch^{k-m+1}\ \|u\|_{k+1,E}.$$

Combining and summing leads to

$$\|u - U_h\|_{m,\Omega} \leq Ch^{1+k-m}\ \|u\|_{k+1,\Omega}\ . \tag{4.4}$$

This result assumes that $\|u\|_{k+1,\Omega}$ exists, this may not be true if the function f is not sufficiently smooth. An example of this lack of smoothness appears in the auxiliary problems (3.4) and (3.5) introduced in the application of the Aubin-Nitsche lemma. If the solution does not possess the necessary smoothness then the value of k in (4.4) has to be reduced until the right hand side does exist. In particular, if $f \in L_2$ and no additional smoothness is available, then

$$\|u\|_{2m} \leq \|f\|_0$$

and we must replace k+1 in (4.4) by 2m.

For second order problems (m=1) with elements that recover polynomials of degree k,

$$\|u - U_h\|_{1,\Omega} \leq Ch^k\ \|u\|_{k+1,\Omega}. \tag{4.5}$$

In view of the lack of smoothness in the error function e, applying (3.3) we have

$$\|w - W_h\|_{1,\Omega} \leq Ch\ \|w\|_2\ , \tag{4.6}$$

with

$$\|w\|_2 \leq C\|e\|_0 = C$$

where C is a constant independent of h. Combining (4.5) and (4.6) yields

$$\|u - U_h\|_{0,\Omega} \leq Ch^{k+1}\ \|u\|_{k+1,\Omega}. \tag{4.7}$$

L_2 estimates incorporating boundary approximation have been considered by Berger (1973).

5. L_∞-ERROR ESTIMATES

The natural norms for deriving finite element error estimates are Sobolev norms $\|\cdot\|_{m,p,\Omega}$, $1 < p < \infty$, the proofs of estimates in terms of the L_∞-norm tend to be highly technical and are derived via Sobolev norm estimates. Optimal L_∞-estimates have been derived using weighted Sobolev norms

$$\|u\|_{m,\Omega;w} = \left\{ \sum_{|k| \leq m} \int_\Omega |D^k u|^2 \; w \; d\Omega \right\}^{\frac{1}{2}} , \tag{5.1}$$

A number of different weight functions w have been used. If $\underline{x}_0$ is a reference point and

$$r \equiv |\underline{x} - \underline{x}_0|$$

then one popular weight function is

$$w(\underline{x}) = (\rho^2 + r^2)^{-\alpha} \qquad \alpha, \rho > 0$$

(Nitsche, 1975, 1976; Natterer, 1977; Ciarlet, 1978). The parameters α and ρ become important after applying Sobolev's lemma, as bounds in terms of weighted Sobolev norms lead to estimates of the form

$$|w(\underline{x}_0)(u - U_h)(\underline{x}_0)| = O(h^k) .$$

If $\rho = O(h)$ then this bound becomes

$$|(u - U_h)(\underline{x}_0)| = O(h^{k+2\alpha}).$$

The weighted norms are only equivalent to the standard forms if restrictions are placed on the values of the parameter; thus the proof is only valid for

$$0 < \alpha < \tfrac{1}{2} \qquad k = 1$$

and

$$0 < \alpha \leq \tfrac{1}{2} \qquad k \geq 2$$

These restrictions lead to the error estimates

$$\|u - U_h\|_{0,\infty,\Omega} \leq C|u|_{k+1,\infty,\Omega} \begin{cases} h^{2-\varepsilon} & k = 1 \\ h^{k+1} & k \geq 2 \end{cases} \tag{5.2}$$

Alternative weighted norms have been introduced (Natterer, 1975; Babuška and Rosenweig, 1972), but they all lead to the same estimates as does a completely different style of proof based on an analysis of the Green's function (Scott, 1976). A reformulation of the term $h^{2-\varepsilon}$ as $h^2|\ell n(h)|^{\frac{1}{2}}$ emphasises the dependence on the Green's function and examples have been constructed (Fried, 1980; Jespersen, 1978) to show that the result is the best possible and not a product of the style of proof. Rannacher and Scott (1982) have shown that these logarithmic factors do not appear in other L_p norms. Many of the results for L_∞-bounds are surveyed in Nitsche (1978) and they have been extended to accommodate numerical integration (Wahlbin, 1978). L_∞-estimates have a number of serious flaws, when formulated as above, they are seriously affected by the lack of smoothness in the neighbourhood of the boundary singularities and they fail to identify the superconvergence properties of the finite element solution. The problem of boundary singularities is covered in a later chapter but it should be noted here that a number of authors (Schatz and Wahlbin, 1977, 1978, 1979, 1981, 1982; Schatz, 1980; Bramble and Thomée, 1974) have isolated this boundary effect and shown that the optimum rate is recovered in the interior of Ω away from the boundary singularities. These same interior estimates also apply when the boundary conditions, or even the boundary itself, is approximated (Schatz and Wahlbin, 1977; Bramble and Thomée, 1974, Nitsche and Schatz, 1974).

Superconvergence is a phenomenon that has been observed frequently in engineering finite element computations, it implies the existence of points within each element at which the approximation is more accurate than would appear from the preceding estimates. In one dimensional problems and tensor product elements in higher dimensions, these superconvergence points may be the nodes (Douglas, Dupont and Wheeler, 1974). A typical result is that the error at the nodes is $O(h^{k+2})$, $k \geqslant 3$ rather than the $O(h^{k+1})$ expected from (5.2). In most engineering calculations, it is the stresses rather than the displacements that exhibit superconvergence and it has been shown (Lesaint and Zlámal, 1979; Zlámal, 1977) that the gradients of the finite element approximation exhibit superconvergence at the Gauss-Legendre points of curved isoparametric elements, provided that they are not too curved. Thus, if $|u - U_h|_{1,2,\Omega}$ is approximated by computing the values at the Legendre-Gauss points only, we have an error in terms of $O(h^{k+1})$ rather than the $O(h^k)$ expected from (4.5).

An alternative form of superconvergence is the technique of optimal recovery suggested by Bramble and Schatz (1974, 1977), another is given by Dupont (1976), in which local averaging is used to compute a modification of the Galerkin solution with $O(h^{k+1})$ accuracy to give an easily obtainable approximation with $O(h^{2k})$ accuracy (see also chapter 6 of this volume).

REFERENCES

Babuška, I. and Rosenweig, M.B. (1972) A finite element scheme for domains with corners. *Numer. Math.*, **20**, 1-21.

Berger, A. (1973) L_2 error estimates for finite elements with interpolated boundary condition. *Numer. Math.*, **21**, 345-349.

Bramble, J.H. and Schatz, A.H. (1974) Higher order local accuracy by averaging in the finite element method. In "Mathematical Aspects of Finite Elements in Partial Differential Equations", (C. de Boor, Ed.), New York, Academic Press.

Bramble, J.H. and Schatz, A.H. (1977) Higher order local accuracy by averaging in the finite element method. *Math. Comp.*, **31**, 94-111.

Bramble, J.H. and Thomée, V. (1974) Interior maximum norm estimates for some simple finite element methods. *R.A.I.R.O. Rouge*, **8**, 5-18.

Ciarlet, P.G. (1978) The Finite Element Method for Elliptic Problems. Amsterdam, North-Holland.

Ciarlet, P.G. and Raviart, P.A. (1972) General Lagrange and Hermite interpolation in R^n with applications to finite element methods. *Arch. Rat. Mech. Anal.*, **46**, 177-199.

Douglas, J., Dupont, T. and Wheeler, M.F. (1974) An L_∞ estimate and a superconvergence result for a Galerkin method for elliptic equations based on tensor products of piecewise polynomials. *R.A.I.R.O. Rouge*, **8**, 61-66.

Dupont, T. (1976) A unified theory of superconvergence for Galerkin methods for two-point boundary problems. *SIAM J. Numer. Analysis*, **13**, 362-368.

Fried, I. (1980) On the optimality of the pointwise accuracy of the finite element solution. *Int. J. Num. Methods Engng.*, **15**, 451-456.

Jespersen, D. (1978) Ritz-Galerkin methods for singular boundary value problems. *SIAM J. Num. Analysis*, **15**, 813-834.

Lesaint, P. and Zlámal, M. (1979) Superconvergence of the gradient of finite element solutions. *R.A.I.R.O. (Num. Anal.)*, **13**, 139-166.

Mitchell, A.R. and Wait, R. (1984) The Finite Element Method in Partial Differential Equations, Chichester, Wiley.

Natterer, F. (1975) Über die punktweise Convergenz Finiter Elemente. *Numer. Math.*, **25**, 67-77.

Natterer, F. (1977) Uniform convergence of Galerkin's method for splines on highly nonuniform meshes. *Math. Comp.*, **31**, 457-468.

Nitsche, J.A. (1975) L_∞-convergence of finite element approximation. In Proc. 2nd Conference on Finite Elements, Rennes.

Nitsche, J.A. (1976) L_∞-convergence of finite element approximations. In "Mathematical Aspects of Finite Element Methods", (I. Galligani and E. Magenes, Eds.), Lecture Notes 606, Berlin, Springer.

Nitsche, J.A. (1978) L_∞-error analysis for finite elements. In "The Mathematics of Finite Elements and Applications", (J.R. Whiteman, Ed.), London, Academic Press.

Nitsche, J.A. and Schatz, A.H. (1974) Interior estimates for Ritz-Galerkin methods. *Math. Comp.*, **28**, 937-958.

Rannacher, R. and Scott, R. (1982) Some optimal error estimates for piecewise linear finite element approximations. *Math. Comp.*, **38**, 437-445.

Schatz, A.H. (1980) A weak discrete maximum principle and stability of the finite element method in L_∞ on plane polygonal domains. Part 1. *Math. Comp.*, **34**, 77-91.

Schatz, A.H. and Wahlbin, L.B. (1977) Interior maximum norm estimates for finite element methods. *Math. Comp.*, **31**, 414-442.

Schatz, A.H. and Wahlbin, L.B. (1978) Maximum norm estimates in the finite element method on plane polygonal domains. Part 1. *Math. Comp.*, **32**, 73-109.

Schatz, A.H. and Wahlbin, L.B. (1979) Maximum norm estimates in the finite element method on plane polygonal domains. Part 2, refinements. *Math. Comp.*, **33**, 465-492.

Schatz, A.H. and Wahlbin, L.B. (1981) On a local asymptotic error estimate in finite elements and its use: Numerical examples. In "Advances in Computer Methods for Partial Differential Equations, IV", (R. Vichnevetsky and R.S. Stepleman, Eds.), New Brunswick, IMACS.

Schatz, A.H. and Wahlbin, L.B. (1982) On the quasi-optimality in L_∞ of the $\overset{\circ}{H}_1$-projection into finite element spaces. *Math. Comp.*, **38**, 1-22.

Schultz, M.H. (1971) L_2 error bounds for the Rayleigh-Ritz-Galerkin method. *SIAM J. Numer. Analysis*, **8**, 737-748.

Scott, R. (1976) Optimal L_∞ estimates for the finite element method on irregular meshes. *Math. Comp.*, **30**, 681-697.

Showalter, R.E. (1977) Hilbert Space Methods for Partial Differential Equations. London, Pitman.

Stakgold, I. (1979) Green's Functions and Boundary Value Problems. New York, Wiley.

Wahlbin, L.B. (1978) Maximum norm error estimates in the finite element method with isoparametric quadratic elements and numerical integration. *R.A.I.R.O. (Num. Anal.)*, **12**, 173-202.

Zlámal, M. (1977) Some superconvergence results in the finite element method. In "Mathematical Aspects of Finite Element Methods", (I. Galligani and E. Magenes, Eds.), Lecture notes 606, Berlin, Springer.

A SHORT SURVEY OF PARABOLIC GALERKIN METHODS

T.F. Dupont

(Department of Mathematics, University of Chicago, USA)

1. INTRODUCTION

The purpose of this paper is to touch on a few topics in that part of the theory of parabolic Galerkin methods related to numerical approximation based on finite element function spaces. It is intended to give a flavour of the field, not to be a comprehensive survey.

In section 2 a model problem is introduced to serve as a basis for the following discussions. A short treatment of some L^2-based theory for the differential problem is given for contrast with the later remarks on the Galerkin theory. In particular, two energy estimates are presented and then a parabolic smoothing result that shows how solutions of parabolic problems gain smoothness in time is outlined.

Next, in section 3, elementary cases for the Galerkin method are presented. Both continuous-time and discrete-time methods are formulated. Three symmetric, or quasi-optimal, error estimates are stated, and a standard type of optimal order L^2 estimate is sketched. A few simple remarks about nonlinear problems are given.

Then, in section 4, several more sophisticated topics are exposed. Specifically, so-called efficient time discretisations in which the work to compute the solution is almost proportional to the number of parameters used to define the solution are discussed. Also some superconvergence and parabolic smoothing results are hinted at.

Finally, in section 5, a short discussion is given of some current research and of my guesses concerning the future development of this area.

2. A MODEL PROBLEM

Let $J = (0,T]$ be the time interval of interest, where T is positive. Take Ω to be a bounded domain in R^d where d is a positive integer. Assume that Ω has a smooth boundary $\partial\Omega$ and denote by ν the outward unit normal to $\partial\Omega$. In the following model problem we are trying to find a function $u(x,t)$ defined for the spatial variable x in Ω and for the temporal variable t in the closure of the interval J. This function will be constrained to satisfy the partial differential equation

$$u_t - \nabla\cdot(a(x)\nabla u + bu) + u = f,\ x \in \Omega,\ t \in J, \qquad (2.1)$$

In this equation the subscript t denotes partial differentiation with respect to time and the del, ∇, represents the gradient or divergence operator in the spatial variables, as is appropriate. The diffusion coefficient, a, will be assumed uniformly positive. The function a and the vector field b will be considered to be functions of the spatial variables and will be assumed to be sufficiently differentiable on the closure of Ω. Likewise the forcing term, f, will be taken to be a nice function on the closure of $\Omega \times J$.

In addition to the differential equation, the function u will be required to satisfy the following conditions on the lateral boundary, $\partial\,\Omega \times J$, and on the initial surface:

$$a \frac{\partial u}{\partial \nu} = 0, \; x \in \partial\,\Omega, \; t \in J, \tag{2.2}$$

$$u(x,0) \text{ given.} \tag{2.3}$$

2.1 A Weak Form of the Problem

The Galerkin methods are closely related to a weak form of this model problem. Let$(\cdot,\cdot)$ be the $L^2(\Omega)$ innerproduct:

$$(v,w) = \int_\Omega vw\, dx.$$

Denote by $H^j(\Omega)$ the Sobolev space of functions having derivatives through order j in $L^2(\Omega)$. (See, for example, other articles in this book or Adams (1975) for a discussion of these spaces.)

As seen in the previous articles we will use bilinear forms to specify the weak form of the problem. Define the bilinear forms A, A_0, and A_1 by the relations

$$A(v,w) = A_1(v,w) + A_0(v,w), \tag{2.4}$$

$$A_1(v,w) = \int_\Omega (a\nabla v \cdot \nabla w + vw)\, dx, \tag{2.5}$$

$$A_0(v,w) = \int_\Omega w\nabla\cdot(b\; v)\, dx. \tag{2.6}$$

The solution to the weak problem is a map of the interval J into the space $H^1(\Omega)$ which is differentiable considered as a map into the dual space $H^{-1}(\Omega)$ of $H^1(\Omega)$ and which satisfies the following relation:

$$(u_t,v) + A(u,v) = (f,v), \quad v \in H^1(\Omega), \; t \in J. \tag{2.7}$$

In addition the solution of this problem is required to converge, in $H^{-1}(\Omega)$, to the given initial data as t goes to zero.

2.2 Two Basic Estimates for the Differential Problem

Perhaps the most elementary bound for the solution of (2.7) is derived by taking v = u at each t. Let $\|\cdot\|$ denote the $L^2(\Omega)$ norm and let $\|\cdot\|_j$ denote the norm on $H^j(\Omega)$. Then, one has the three relations for some positive α

$$(u_t,u) = \frac{1}{2}\frac{d}{dt}(\|u\|^2),$$

$$A_1(u,u) \geq \alpha \|u\|_1^2,$$

$$|A_0(u,u)| \leq C \|u\| \|u\|_1.$$

These together with (2.7) give

$$\frac{d}{dt}\|u\|^2 + 2\alpha \|u\|_1^2 \leq C\|u\|\|u\|_1 + 2\|f\|_{-1}\|u\|_1. \tag{2.8}$$

where the $H^{-1}(\Omega)$- norm of f is the norm it has as an element of the dual space of $H^1(\Omega)$.

From (2.8) it easily follows that

$$\frac{d}{dt}\|u\|^2 + \alpha \|u\|_1^2 \leq C[\|u\|^2 + \|f\|_{-1}^2]. \tag{2.9}$$

This in turn implies, by Gronwall's lemma, that there is a C such that

$$\|u\|^2_{L^\infty(J;L^2)} + \|u\|^2_{L^2(J;H^1)} \leq C[\|u(0)\|^2 + \|f\|^2_{L^2(J;H^{-1})}], \tag{2.10}$$

where the $L^p(J;H^j)$-norm is the L^p norm on J of the $H^j(\Omega)$-norm of the function. This bound, (2.10), is the first basic "energy" estimate for the solution of the differential problem.

A second energy estimate for the solution of the differential problem is obtained by using $v = u_t$ in (2.7). Using manipulations similar to those above one gets

$$\|u\|^2_{L^2((s,t);L^2)} + \|u(t)\|_1^2 \leq C[\|u(s)\|_1^2 + \|f\|^2_{L^2((s,t);L^2)}]. \tag{2.11}$$

The key estimate in deriving this relation is

$$A_1(u,u_t) = \frac{1}{2}\frac{d}{dt}A_1(u,u). \tag{2.12}$$

To get higher-order bounds one can differentiate the equation (2.7) with respect to t and repeat the two arguments used to get the estimates (2.10) and (2.11).

Notice that the bounds presented in this section are all based on L^2 theory. There are of course many other types of bounds for parabolic equations, but the L^2-based results were used here since they go over most naturally to estimates for Galerkin approximations.

2.3 Parabolic Smoothing

Take $f \equiv 0$. Under this condition the solution u gets smoother in time in a way that is easy to quantify.

Integrate (2.11) from $s = 0$ to $s = t$ to see that

$$t \|u(t)\|_1^2 \le C \|u\|^2_{L^2((0,t);H^1)}. \tag{2.13}$$

Now use (2.10) to see that

$$t \|u(t)\|_1^2 \le C \|u(0)\|^2.$$

Hence,

$$\|u(t)\|_1 \le C\, t^{-1/2} \|u(0)\|. \tag{2.14}$$

In general, with a, b and Ω sufficiently nice and $f \equiv 0$,

$$\|u(t)\|_k \le C\, t^{-k/2} \|u(0)\|. \tag{2.15}$$

3. PARABOLIC GALERKIN METHODS

Galerkin methods can be viewed as methods in which the spatial variable has been discretised or as methods in which the spatial and temporal variables have been discretised. There is some benefit in considering a case in which the temporal variable has not been discretised, because the proofs are usually simpler than, but very similar to, the proofs in the corresponding fully discrete case.

3.1 A Continuous-Time Galerkin Method

Let M be a finite dimensional subspace of $H^1(\Omega)$. Find a differentiable map $U: J \to M$ such that

$$(U_t, V) + A(U,V) = (f,V),\ V \in M,\ t \in J, \tag{3.1}$$

where $U(0)-u(0)$ is small.

3.1.1 A Symmetric Error Bound

The first result for such a method is that given U(0) there is a unique map U. This is true because, given a basis for M, the relations (3.1) are equivalent to a set of linear ordinary differential equations; the coefficient of the time derivative term is the Gramm matrix, hence nonsingular.

In the case of the elliptic problems studied in other articles in this book (and many others) the first type of error estimate produced is what is called a symmetric or quasi-optimal bound. Such estimates say that if the solution can be approximated well by something in the finite dimensional function space M then it will be, and further the norm used is the same in both parts of the proposition. For second order elliptic problems the natural norm is the $H^1(\Omega)$ norm. In the case of the continuous-time Galerkin method (3.1) such a bound could be phrased as follows:

For some norm $|||\cdot|||$ there is a constant C such that for all M

$$||| U-u ||| \leq C \inf\{||| V-u ||| : V \text{ in } C^1(\bar{J};M)\}. \qquad (3.2)$$

This says that in the particular norm that the Galerkin approximation is, up to a constant multiple, the best possible approximation.

There are three norms for which such a result is known, each with some restrictions. These are the norms

$$||| V ||| = \|V\|_{L^\infty(L^2)} + \|V\|_{L^2(H^1)} + \|V_t\|_{L^2(H^{-1})},$$

$$||| V ||| = \|V\|_{L^\infty(L^\infty)},$$

$$||| V ||| = \|V\|_{L^2(H^1)} + \|D_t^{1/2}V\|_{L^2(L^2)}.$$

These results are examined in Dupont (1982), Nitsche and Wheeler (1982) and Douglas and Dupont (1970) respectively. In this regard one should also see the result of Thomée and Wahlbin (1983).

3.1.2 An L^2 Error Bound

The discussion that follows is a simple example of a type of argument that is frequently used to get optimal-order $L^2(\Omega)$ error bounds for parabolic Galerkin methods. This type of argument is due to Wheeler (1973) and has been refined by many people. The particular details here resemble most closely those in Dupont (1972).

Define a map W: J → M, which we cannot compute because we do not know u, by

$$A_1(W,V) = A_1(u,V), \quad V \in M, \quad t \in J. \qquad (3.3)$$

Then, at each time t, W is a Galerkin approximation to the solution of an elliptic problem which has u as its solution. Hence we know by the well-understood theory for elliptic problems that W is a good approximation of u in the $L^2(\Omega)$-norm.

Let $Z = U-W$ and $Q = u-W$. For simplicity take A_0 to vanish identically. Then

$$\begin{aligned}(Z_t,V) + A(Z,V) &= (Q_t,V) + A(Q,V) \\ &= (Q_t,V), \quad V \in M,\end{aligned} \tag{3.4}$$

where the last step used the definition of W to remove the A form from the right hand side. With V = Z one gets, just as in section 2, that

$$\|Z\|_{L^\infty(J;L^2)} + \|Z\|_{L^2(J;H^1)} \le C[\|Z(0)\| + \|Q_t\|_{L^2(J;H^{-1})}], \tag{3.5}$$

note that since Z is in M it is legitimate to use V = Z in (3.4).

If, say, U(0) is the $L^2(\Omega)$ projection of u(0) into M, it then follows that for all t in J

$$\|U(t) - u(t)\| \le C[\|Q\|_{L^\infty(J;L^2)} + \|Q_t\|_{L^2(J;H^{-1})}]. \tag{3.6}$$

If M consists of piecewise polynomials of degree at most r over a nice mesh of size h, and if $f \equiv 0$, and if u(0) satisfies certain compatibility conditions, then (3.6) can be translated into

$$\|U-u\|_{L^\infty(J;L^2)} \le C\, h^{r+1}\|u(0)\|_{r+1} . \tag{3.7}$$

3.2 Discrete-Time Galerkin Methods

The procedure (3.1) is not a computational process since it involves the solution of a set of ordinary differential equations, and, for all but the simplest cases, they cannot be solved analytically. What is usually done is to discretise time and produce a finite difference version of (3.1). It might seem strange that one would not use finite elements in time, and that can be done; however, in time the equations are first order and the most natural finite element methods for first order equations are not very good. Because of this, most of the literature on fully-discrete parabolic Galerkin methods involves finite differences in time.

Let the interval J be partitioned into N pieces by the partition $0 = t_0 < t_1 < \dots < t_N = T$. Denote by U^n the fully-discrete approximation to $U(t_n)$. Let $\Delta t_n = t_{n+1} - t_n$ and set $d_t U^n = (U^{n+1} - U^n)/\Delta t_n)$. For θ in the interval [1/2,1] let $U^{n,\theta} = U^n + \theta(U^{n+1} - U^n)$. A so-called θ-weighted approximation to U is defined by

$$(d_t U^n, V) + A(U^{n,\theta}, V) = (f^{n,\theta}, V),\ V \in M, \qquad (3.8)$$

for $n = 0, 1, \ldots, N - 1$.

It is easily seen that, provided $\max \Delta t_n$ is sufficiently small, U^0 and θ uniquely determine the sequence $U^1, \ldots, U^N$.

An error estimate that is almost identical with (3.6) can be obtained for this method, except that the right hand side has a time truncation term in it. The time truncation is on the order of Δt if $\theta \neq 1/2$ and is on the order of Δt^2 if $\theta = 1/2$; here Δt is the maximum of the Δt_n's. See, for example, Dupont (1972) for an explicit representation of the time truncation terms.

As a special case, suppose that $\theta = 1/2$ and that the function space M consists of all piecewise cubic polynomials over a nice mesh of size h. Then the L^2-norm of the error at each t_n is bounded by $C[h^4 + \Delta t^2]$, provided that the solution is sufficiently smooth.

3.3. A Galerkin Method for Nonlinear Parabolic Equations

Much of the theory hinted at above is valid in the case of nonlinear parabolic problems. A large collection of methods was treated in Douglas and Dupont (1970); the analysis there of the methods treated is primitive by comparison with later work, but no later work treats quite so many different processes. The paper by Wheeler (1973) gets optimal $L^2(\Omega)$ error estimates for some nonlinear problems, and some of the references in section 4 are primarily concerned with nonlinear problems.

Assume now that A_0 vanishes identically for simplicity and use the notation of section 3.2. Assume that the diffusion coefficient a is a function of x and u; i.e., the diffusion depends on the solution. Let the bilinear A form now be parametrised by another function:

$$A(z,v,w) = \int_\Omega (a(x,z(x))\ \nabla v \cdot \nabla w + vw)\,dx. \qquad (3.9)$$

Let an extrapolation operator E^n be defined as follows:

$$E^n = U^n + \theta \Delta t d_t U^{n-1} \text{ for } n \geq 1. \qquad (3.10)$$

For $n = 0$ use $E^0 = U^0$. Given U^0, define the sequence $U^1, \ldots, U^N$ by

$$(d_t U^n, V) + A(E^n; U^{n,\theta}, V) = (f^{n,\theta}, V),\ V \in M. \qquad (3.11)$$

This requires the solution of a linear system of equations at each time step even though the differential equation is nonlinear. This particular extrapolation scheme has been analysed as have several others. A particularly nice touch, I think, is using a predictor-corrector method in which the predictor is determined using the corrector equations from the last step [Douglas and Dupont (1970)].

The solution to the parabolic problem must be smooth or there is essentially no analysis available in the nonlinear case. In the linear problem, as we shall see below, this is not so.

4. SOME ADVANCED TOPICS IN PARABOLIC GALERKIN METHODS

The topics addressed in this section are directed at several questions.

How can the finite element approximation be computed more cheaply?

How can the information needed from the approximate solution be extracted more accurately?

How can the estimates of the size of the error be made more reliable?

4.1 Efficient Time Discretisation

The title of this section is a name that is used to describe a class of methods that have been devised to allow the calculation of the solutions to finite element solutions in an amount of work that is bounded by a multiple of the number of parameters involved in the definition of the solution. Some of these methods fall short of the goal because a logarithm is involved in the multiple, but that is close enough to be included in this class.

In finite difference theory the easiest discretisations of the type sought here are the explicit methods, and by a technique called "mass lumping" such methods can be constructed in the finite element context. However, for parabolic problems explicit time differencing leads to time steps that are frequently much too small for the accuracy required; the constraint on the time step comes from stability conditions; this holds for both finite difference and finite element methods.

The first implicit methods to achieve the same type of work requirement per time step as explicit methods were the alternating direction methods invented for finite difference methods [Peaceman and Rachford, (1955)]. These have been studied for finite elements too, and they will be discussed in section 4.2. The methods that are the principal object of this section are intended to work with more generality than can be expected of alternating direction methods.

To provide a contrast to the techniques of this section an operation count for a respectable approach to solving the equations of section 3 is needed. For definiteness assume that we are dealing with the case of a uniform time step of size Δt, that θ is 1/2, that the spatial mesh is on a two dimensional domain, and that the size of the mesh is approximately h. Suppose that we use conjugate gradient iteration to solve the equations approximately at each time step; either take A_0 to vanish or evaluate it at a previous time level.

With the above assumptions the work required to reduce the error in the approximate solution by a factor of Δt to some power is $O(h^{-2}\ h^{-1}\ \ln(1/\Delta t))$. Since the error involves $h^4 + \Delta t^2$, we would take Δt about like h^2 for balance. Thus we would need $O(h^{-2})$ time steps.

The result is that the total work for this approach is $O(h^{-5}\ \ln(h^{-1}))$. The number of parameters that define the solution is $O(h^{-4})$.

A simple example of an efficient time discretisation is as follows: In taking the time step that will produce U^{n+1}, do the following. First, extrapolate to get a good approximation of the solution. Next, reduce the error by a fixed factor and accept the answers as U^{n+1}. One reasonable way to reduce the error by a fixed factor is to use a conjugate gradient iteration that is preconditioned by an exact factorisation done for the first time step.

With this technique the work can be itemised as follows:

Work to factor (done only once) $= O(h^{-3})$.

Work to solve each step $= O(h^{-2}\ln(h^{-1}))$.

Thus the total work is $O(h^{-4}\ln(h^{-1}))$.

There are many variants of this approach discussed in Douglas, Dupont and Percell (1977), Douglas, Dupont and Ewing (1979) and Ewing and Russell (1983). It is directed primarily at nonlinear problems and is perhaps even more useful in three dimensional problems than in two dimensional ones.

4.2 Alternating Direction Methods

In the situation when the domain Ω is a rectangle, or a rectangular parallelopiped, the space M is frequently taken to be a tensor product space. In the case of a rectangle, two finite element spaces consisting of functions of one variable, M_x and M_y, are chosen, then M consists of all finite sums of products $f(x)g(y)$ where f and g are in M_x and M_y, respectively.

To describe an alternating direction method simply we consider the special case of the heat equation; i.e.,

$$A(v,w) = \int_\Omega \nabla v\cdot\nabla w\ dx. \tag{4.1}$$

In this case define the sequence U^n by relations of the form

$$(d_tU^n,V) + \theta\Delta t A(d_tU^n,V) + (\theta\Delta t)^2\int_\Omega d_tU^n_{xy}V_{xy}dxdy$$
$$= -A(U^n,V) \text{ for } V \text{ in } M. \tag{4.2}$$

This has the algebraic form

$$(C_x + \theta\Delta t\, A_x)\otimes(C_y + \theta\Delta t\, A_y)d_t U^n = -A\, U^n. \tag{4.3}$$

Such tensor product operator equations can be solved by working with a collection of problems based solely on one space dimensional problems, and the solution of such collections is cheaper by far than the solution of higher space dimensional problems.

For Galerkin methods these methods were introduced by Douglas and Dupont (1971). Their range of applicability was extended by Dendy and Fairweather (1975), Dendy (1977), and Hayes (1981). In some of this work the constraint that the domain be a rectangle is considerably reduced.

4.3 Superconvergence

The phenomenon of superconvergence occurs when the computed approximation is closer to the true solution at certain points than is possible globally. The concept has also been generalised to include the situations in which a quantity, such as a flux, is computed from the approximate solution that is more accurate than if it were done on a pointwise basis. Superconvergence has been most extensively studied in the context of one-dimensional elliptic problems; see Dupont (1976) and the references therein.

As a first example of superconvergence consider the case of $\Omega = (c,d)$ a one-dimensional interval. Let M consist of all continuous piecewise polynomials of degree at most r over a given mesh of size h. Then under reasonable hypotheses the error at the mesh points is $O(h^{2r})$. Since the best that one can hope for in general is $O(h^{r+1})$ as a bound for the error in the maximum norm, we see that for r at least two the approximation at the mesh points is better than we should expect at other points. For more detail on this type of situation see Thomee (1980) and Douglas, Dupont and Wheeler (1978).

As another instance of this effect consider the case of a rectangle with a tensor product function space which is built from spaces of the type just used in the one-dimensional example. Take r to be at least two and assume that the solution is smooth. Then at the corners of the mesh blocks the error is $O(h^{r+2})$ instead of the $O(h^{r+1})$ one might have expected. Further the approximation of the gradient is one power of h better than is possible globally at the r by r grid of Gauss points on each block. See Douglas, Dupont, Wheeler (1974).

A different aspect of superconvergence can be seen in the technique for flux calculation of John Wheeler [Wheeler, (1973) and (1978)]. Several variants of it have been described and analysed by Wheeler (1974), Dupont (1976) and Douglas, Dupont and Wheeler (1974). This technique is useful in situations in which the boundary values are given, as opposed to the boundary flux in the model problem, and the flux needs to be computed.

If one uses a mesh of the type in the previous example and computes an integral of the flux by evaluating the flux at many points on the boundary and integrating it numerically along the boundary or a part thereof, then the error will in general be $O(h^r)$. If however one uses an auxiliary calculation involving the solution in the interior of the region, a superconvergent approximation with error $O(h^{2r})$ can be produced.

The auxiliary calculation called for in this method is very simple and the tools to do it are usually present in the program that computes the approximate solution. It is worth remarking that this approach also works well in the case of an elliptic problem in which a flux calculation is required.

4.4 Parabolic Smoothing for Galerkin Methods

There is a property of parabolic Galerkin Methods that is very closely linked with the smoothing effect discussed in section 2.3; that is if the initial data for the Galerkin method are given by the L_2 projection into the finite-dimensional function space then even if the initial data are so rough that optimal order convergence is not possible, optimal order convergence can occur at later time.

Suppose the U is the solution of (3.1), that U(O) is the $L^2(\Omega)$ projection of the initial data u(O), that a, b and Ω are all sufficiently smooth, and that $f \equiv 0$. Suppose also that the finite dimensional space M consists of piecewise polynomials of degree at most r over a nice mesh of size h. Then there is a constant C such that

$$\|(u-U)(t)\| \leq C\, h^k t^{-k/2} \|u(0)\|, \; 0 \leq k \leq r+1. \tag{4.4}$$

Thus at any positive time the error goes to zero like h^{r+1}.

This parabolic smoothing property is connected with the negative index Sobolev norm approximation of the initial values. This point is perhaps most clearly made in Thomée (1980).

There are many papers in this subject of varying degrees of complexity. Perhaps the simplest is given in Luskin and Rannacher (1982). In any event, the Luskin and Rannacher paper contains many references and some discussion of the development of this topic. The work on parabolic smoothing is, in my view, some of the most elegant of all the work that has been done on finite element methods.

Some unpublished experimental evidence indicates that this remarkable property does not always hold in the case of nonlinear parabolic equations.

5. CURRENT RESEARCH AREAS

The theory of parabolic Galerkin methods has reached a reasonable degree of maturity, but there are still several open areas of interest.

Those topics in the theory necessary to support the development of robust software will be pursued by many. Some of the areas that need refinement are

a posteriori error estimation,

adaptive meshing in space,

adaptive order in space and time.

These all come under the umbrella of adaptive error and efficiency control. Some recent work that is addressed to these questions can be found in Bieterman and Babuska (1983).

An area that is hardly developed at all, but which has considerable potential in terms of improvements in efficiency, is the construction and analysis of algorithms that use different time steps in different parts of the region. It is frequently the case that disturbances in parabolic problems are introduced in one area or on one side of the spatial domain. In these cases the entire domain needs to be modelled, but the solution is very much more tame in some parts of the region than in others.

Convection dominated problems still pose problems in the sense that they are in between the rather nice parabolic problems and the much tougher (for standard methods) hyperbolic problems. These problems seem to call for some upwinding of one sort or another; this is a topic that is treated in at least one other article in this book. Moving meshes, see Miller and Miller (1981), Miller (1981), Dupont, (1982), can be used on such problems with apparent success, but much is left to be understood about such methods. It seems that blending various methods may well prove useful in treating these problems.

The impact of vector machines on the future of calculation is something which is poorly understood at this point. It is clear that the true costs of various methods will depend on how they are adapted to the available hardware. At least initially, it is likely that highly structured hardware will give an advantage to algorithms that are more regular. There are, however, adaptive methods that from the point of view of the data structure are highly regular, so I do not expect that all future work will be done on uniform grids by lightning fast vector processors.

REFERENCES

Adams, R.A. (1975) Sobolev Spaces, Academic Press, New York

Bieterman, M. and Babuska, I. (1983) The finite element method for parabolic equations I and II, *Numer. Math.*, **40**, p. 339

Dendy J.E. (1977) An alternating direction method for Schrodinger's equation, *SIAM Jour. Numer. Anal.*, **14**, p. 1028

Dendy, J.E. and Fairweather G. (1975) Alternating-direction Galerkin methods for parabolic and hyperbolic problems on rectangular polygons, *SIAM Jour. Numer. Anal.*, **12**, p. 144

Douglas, J., Jr. and Dupont, T. (1970) Galerkin methods for parabolic equations, *SIAM Jour. Numer. Anal.*, **7**, p. 575

Douglas, J., Jr. and Dupont, T. (1971) Alternating-direction Galerkin methods on rectangles, Numerical Solution of Partial Differential Equations-II, (Hubbard ed.), Academic Press, New York, p. 133

Douglas, J., Jr., Dupont, T and Wheeler, M.F. (1974) A Galerkin procedure for approximating the flux on the boundary for elliptic and parabolic boundary value problems, *RAIRO Numer. Anal.*, **8**, p. 61

Douglas, J., Jr., Dupont, T. and Wheeler, M.F. (1978) A quasi-projection analysis of Galerkin methods for parabolic and hyperbolic equations, *Math. Comp.*, **32**, p. 345

Douglas, J., Dupont, T. and Percell P. (1977) A timestepping method for Galerkin approximations for nonlinear parabolic equations, Lecture Notes in Mathematics 630, Springer Verlag, p. 64

Douglas J., Jr., Dupont, T. and Ewing R.E. (1979) Incomplete iteration for time-stepping a Galerkin method for a quasilinear parabolic problem, *SIAM Jour. Numer. Anal.*, **16**, p. 503

Dupont, T. (1972) Some L^2 error estimates for parabolic Galerkin methods, The Mathematical Foundations of the Finite Element Method with Applications to Partial Differential Equations, (A.K. Aziz, ed), p. 491

Dupont, T. (1976) A unified theory of superconvergence for Galerkin methods for two point boundary value problems, *SIAM Jour. Numer. Anal.*, **13**, p. 362

Dupont, T. (1982) Mesh modification for evolution equations, *Math. Comp.*, **39**, p. 85

Ewing, R.E. and Russell T.F. (1982) Efficient time-stepping procedures for miscible displacement problems, *SIAM Jour. Numer. Anal.*, **19**, p. 1

Hayes, L. (1981) Galerkin alternating-direction methods for nonrectangular regions using patch approximations, *SIAM Jour. Numer. Anal.*, **18**, p. 127

Luskin M. and Rannacher R. (1982) On the smoothing property of the Galerkin method for parabolic equations, *SIAM Jour. Numer. Anal.*, **19**, p. 93

Miller K. (1981) Moving finite elements, part II, *SIAM Jour. Numer. Anal.*, **18**, p. 1033

Miller, R. and Miller K. (1981) Moving finite elements, part I, *SIAM Jour. Numer. Anal.*, **18**, p. 1019

Nitsche J. and Wheeler, M.F. (1982) The L^∞ boundedness of the finite element Galerkin operator for parabolic problems, *Numer. Funct. Anal. and Opt.*, **4**, p. 325

Peaceman D.W. and Rachford H.H. (1955) The numerical solution of parabolic and elliptic partial differential equations, *J. Soc. Ind. Appl. Math.*, **3**, p. 28

Thomée, V. (1980) Negative norm estimates and superconvergence in Galerkin methods for parabolic equations, *Math. Comp.*, **34**, p. 93

Thomée, V. and Wahlbin, L.B. (1983) Maximum norm stability and error estimates in Galerkin methods for parabolic equations in one space variable, *Numer. Math.*, **41**, p. 345.

Wheeler, J.A. (1973) Simulation of heat transfer from a warm pipeline buried in permafrost, AICHE paper 27b 74th National Meeting, 1973

Wheeler, J.A. (1978) Permafrost thermal design for the trans-Alaska pipeline, Moving Boundary Problems, (D.G. Wilson et al, eds), Academic Press, New York, p. 267

Wheeler, M.F. (1973) A priori L^2 error estimates for Galerkin approximations to parabolic partial differential equations, *SIAM Jour. Numer. Anal.*, **10**, p. 723

Wheeler, M.F. (1974) A Galerkin procedure for estimating the flux for a two point boundary problem, *SIAM Jour. Numer. Anal.*, **11**, p. 764

NONCONFORMING ELEMENTS

D.F. Griffiths and A.R. Mitchell

(Department of Mathematical Sciences, University of Dundee)

1. INTRODUCTION

Classical, or conforming, finite element methods for solving elliptic partial differential equations are based on variational principles. The problem is first cast in variational form in which, typically, a quadratic functional (referred to as the 'energy' of the system) has to be minimised over a specified class of functions, say H. Functions in H must possess certain smoothness properties as well as meeting the so-called essential boundary conditions of the problem. An approximation scheme is established by constructing (modulo boundary conditions) a family of subspaces $S^h \subset H$, characterised by a parameter h which tends to zero, and minimising, for each h, the energy over the finite dimensional subspace. Thus, for each fixed h, we have the desirable property that the approximate solution is in fact the best approximation to the exact solution measured in the energy norm. Moreover, in terms of the dictum 'consistency + stability = convergence', consistency can be ensured by endowing the family S^h with sufficient approximation power, stability is a consequence of the inclusion $S^h \subset H$ (positive definiteness of the energy on S^h follows automatically if it is positive definite on the larger space H) and the family of approximations must therefore converge as $h \to 0$.

Given this idyllic situation, it is not unnatural to ask why it should be necessary to perturb it. Two reasons are often advanced to justify the introduction of nonconforming methods, i.e., those methods for which $S^h \not\subset H$. The most obvious reason is that it may be inefficient to construct S^h to meet the smoothness requirements of H. For instance, quintic polynomials are needed to generate a continuously differentiable piecewise polynomial interpolant on a general grid of triangular elements. Secondly, because the energy of the discrete system is, of necessity, higher than the true energy, the approximation might be said to be too stiff and the family of discrete solutions converges from one side. Yet another situation where nonconforming methods may be used to advantage is that where a side constraint, such as the incompressibility condition $\nabla \cdot u = 0$, is imposed. By relaxing the continuity requirements, one can, for piecewise polynomials of fixed degree, obtain an interpolant containing more degrees of freedom and thereby ease the problems of meeting the constraint, at least approximately.

Once nonconforming approximations are admitted, the question of convergence has to be re-examined. Though convergence estimates do give a measure of security when the grid size h is sufficiently small, they are usually of more interest to the analyst than the practitioner; the latter has to perform computations with a fixed, finite value for h.

(We do not wish to give the impression that convergence results are irrelevant, merely that they give only part of the overall picture.) One of the most successful attempts at formulating conditions under which nonconforming approximations are admissible is given by the 'patch test' of Irons (Bazeley et al, 1965). Based on physical arguments, it requires that the approximation be capable of exactly reproducing certain states of constant stress. The patch test was subsequently rephrased in more mathematical terms by Strang (see Strang and Fix, 1973); we shall describe the precise form of this test in the next section. Though the case for the patch test is persuasively argued by Strang and Fix, a precise link with convergence could not be established. The numerical experiments of Sander and Beckers (1977) show that the patch test is not necessary except, perhaps, in the limit $h \to 0$ whilst the examples of Stummel (1980) (see Section 4) cast doubt on its ability to provide sufficient conditions for convergence. Difficulties in interpreting these results are exacerbated because of a lack of a precise definition of what constitutes a finite element or a patch. It is almost impossible, therefore, to gauge the extent to which Stummel's examples are pathological. In section 5 we describe a generalised form of the patch test due to Stummel (1979) which overcomes the mathematical problems by providing both necessary and sufficient conditions for convergence.

2. THE ANALYSIS OF NONCONFORMING METHODS

To illustrate the principal features of the analysis of nonconforming methods applied to elliptic boundary value problems of degree 2m, we concentrate on two model equations. For m = 1 we consider the Poisson equation

$$-\nabla^2 u = f \tag{2.1a}$$

and, for m = 2, the inhomogeneous biharmonic equation

$$\nabla^4 u = f. \tag{2.2a}$$

Both equations are assumed to hold a bounded polygonal domain Ω in $\mathbb{R}^2$ with boundary $\partial\Omega$ on which we impose the Dirichlet conditions

$$u = g_0 \tag{2.1b}$$

for (2.1a) and

$$u = g_0, \quad \frac{\partial u}{\partial \nu} = g_1 \tag{2.2b}$$

in the case of (2.2a), where ν denotes the outward normal to $\partial\Omega$. Let g be any smooth function (for instance $g \in H^m(\Omega) \cap C^0(\bar{\Omega})$) which matches the boundary data $g = g_0$ and, in the case m = 2, $\partial g/\partial\nu = g_1$ on $\partial\Omega$. We may then write the above problems in standard variational form: Find $u - g \in H_0^m(\Omega)$ such that

$$a(v,u) = (v,f) \quad \forall v \in H_0^m(\Omega) \tag{2.3}$$

where $(\cdot,\cdot)$ denotes the usual $L_2(\Omega)$ inner product and the bilinear form $a(\cdot,\cdot)$ is defined by

$$a(v,u) \equiv \int_\Omega \nabla v \cdot \nabla u \, d\Omega \tag{2.4}$$

for m = 1 and, when m = 2,

$$a(v,u) \equiv \int_\Omega \nabla^2 v \, \nabla^2 u \, d\Omega. \tag{2.5}$$

The finite element approximation of (2.3) proceeds by dividing Ω into non-overlapping elements K - with a notional diameter h - on which is defined a space S^h of functions whose restrictions to each element are polynomials of degree, say, r. For a conforming method we have $S^h \subset H^m(\Omega)$ and this implies (see Ciarlet, 1978) that S^h must belong to the class C^{m-1} of functions which are m-1 times continuously differentiable on the closure of Ω. This property is stronger than that implied by the Sobolev embedding results because the piecewise polynomial structure of S^h precludes any form of logarithmic singularity.

Without the inclusion $S^h \subset H^m(\Omega)$, the method becomes one of non-conforming type, the principal effect of which is that the bilinear form $a(\cdot,\cdot)$ is no longer defined on $S^h \times S^h$ because the integrands in (2.4) and (2.5) will involve squares of δ-functions. This difficulty is circumvented in a natural way by replacing integrals over Ω by the sum of integrals over elements. Instead of (2.4), we therefore have

$$a_*(v,u) = \sum_{K \in \Omega} \int_K \nabla v \cdot \nabla u \, d\Omega. \tag{2.6}$$

Some provision must be made for satisfying, at least approximately, the essential boundary conditions associated with each of the problems. Restricting ourselves to nodal finite elements (i.e. those for which $U \in S^h$ means that $U|_K$ is uniquely defined by its values and possibly derivatives at a given set of nodes on K), let $G \in S^h$ denote the interpolant of g so that, for $U \in S^h$, we have $U - G \in S_0^h$, the space obtained by interpolating homogeneous boundary conditions on $\partial\Omega$. A similar provision can be made for more general types of element.

We shall postpone a discussion of the biharmonic problem until later in this section and concentrate meanwhile on the case m = 1. Using (2.6), we define a discrete energy norm on S_0^h by

$$|V|_* = \{a_*(V,V)\}^{1/2} \tag{2.7}$$

and, to avoid the possibility that $|V|_* = 0$ for $0 \neq V \in S_0^h$, we insist that the degree of the underlying polynomial space satisfy $r \geq m$. The nonconforming finite element approximation of (2.3) now reads: Find $U - G \in S_0^h$ such that

$$a_*(V,U) = (V,f) \quad \forall V \in S_0^h. \tag{2.8}$$

It is now natural to enquire as to the conditions under which the solution U of (2.8) converges to that of (2.3) as $h \to 0$ with respect to the norm (2.7). Broadly following Strang and Fix (1973), we can write

$$a_*(V,u-U) = a_*(V,u) - (V,f).$$

When $a_*(\cdot,\cdot)$ is continuous on $S_0^h \times S_0^h$, application of the Cauchy-Schwartz inequality gives

$$C|u-U|_*|V|_* \geq |a_*(V,u) - (V,f)|$$

and hence

$$|u-U|_* \geq C^{-1}\Delta(u) \tag{2.9}$$

where

$$\Delta(u) = \max_{V \in S_0^h} \frac{|a_*(V,u) - (V,f)|}{|V|_*} \tag{2.10}$$

The inequality (2.9) provides a lower bound on the error; the quantity $\Delta(u)$ is also involved in the upper bound. Rearranging the identity

$$|V+G-U|_*^2 = a_*(V+G-U,V+G-U)$$

we find, with $W = V+G-U \in S_0^h$,

$$|W|_*^2 = a_*(V+g-u,W) - a_*(g-G,W) + a_*(W,u) - (W,f)$$

$$\leq C|V+g-u|_*|W|_* + C|g-G|_*|W|_* + |a_*(W,u) - (W,f)|.$$

Hence, dividing both sides by $|W|_*$ and taking the maximum over $W \in S_0^h$, we find,

$$|W|_* \le C\{|V+g-u|_* + |g-G|_* + \Delta(u)\}.$$

Consequently, by the triangle inequality,

$$\begin{aligned} |u-U|_* &= |(u-g-V) + (g-G) + (V+G-U)|_* \\ &\le |V+g-u|_* + |g-G|_* + |W|_* \\ &\le C\{\inf_{V \in S_0^h} |V+g-u|_* + |g-G|_* + \Delta(u)\}. \end{aligned} \tag{2.11}$$

(We denote by C a generic constant that does not necessarily have the value on successive appearances.) The first two terms on the right of (2.11) represent the error in approximating u and g by elements of S^h; we shall assume that S^h has sufficient approximating power to ensure that these terms vanish as $h \to 0$. The remaining term $\Delta(u)$ provides a measure of the nonconformity of the method. For a conforming method we would have $V \in S_0^h \subset H_0^1(\Omega)$, $a_* \equiv a$ and hence $\Delta(u) = 0$ by virtue of the fact that we could choose $v = V$ in (2.3). The convergence of nonconforming methods therefore depends on whether $\Delta(u) \to 0$ as $h \to 0$ for each choice of S_0^h. In the case of the Poisson problem, we find

$$a_*(V,u) - (V,f) = \sum_{K\in\Omega} \int_K \nabla V\cdot\nabla u \, dK + \int_\Omega \nabla^2 u \, V \, d\Omega$$

$$= \sum_{K\in\Omega} \int_{\partial K} V \frac{\partial u}{\partial \nu} \, ds \quad \text{(by Green's Theorem)} \tag{2.12}$$

$$= \sum_{\text{edge}} \int_{\text{edge}} [V] \frac{\partial u}{\partial \nu} \, ds \tag{2.13}$$

where $\partial/\partial\nu$ is the outward normal derivative on the element boundary ∂K and the summation in (2.13) extends over all edges of the grid. We have also written [V] to represent the jump in V across an edge and assumed that $V \equiv 0$ outside Ω. The forms (2.12) and (2.13) are each useful in different situations and examples will be presented in the next section.

The estimation of $\Delta(u)$ is a non-trivial task in general and it would be convenient if some simple criterion were available which would be capable of distinguishing those elements which lead to convergent approximations from those which do not. Borrowing from the ideas of finite differences, we look for a property analogous to consistency for then, by the Lax Equivalence Theorem, stability (in the form of ellipticity on S_0^h) would then be necessary and sufficient for convergence. However, irregular grids and the use of derivatives as well as function values as parameterisations of S_0^h lead to a complexity not normally encountered in the analysis of finite difference methods. In 1965 Irons proposed the patch test as a means of providing such a criterion. Principally, the patch test requires that, if the conditions on the boundary of a patch of elements are chosen to be compatible with a solution whose mth derivatives are constant, then the solution of the finite element equations on this patch should reproduce this presumed solution exactly. Thus, for an elliptic equation of order 2m, if the data (source term and boundary conditions) are modified so that the solution u of this belongs to P_m, the space of polynomials of degree m, the solution U of the finite element equations on this patch should give $U \equiv u$ if the test is to be satisfied. Naturally, for $U \in S^h$, the restriction of U to any element should contain complete polynomials of degree m. The exact reproduction of the solution $u \in P_m$ is convenient in practice because the test may then be applied numerically via a computer program. On theoretical grounds, one could relax the condition $U \equiv u$ to read $U \to u$ as $h \to 0$ but this would pose considerable practical problems for irregular grids since it would require specification of a strategy for mesh subdivision.

We have thus far omitted to define what is meant by a patch and hesitate to be precise. Roughly speaking, a patch is a collection of elements large enough to contain all the representative types of nodes, for example, vertex and midside nodes. The number of elements meeting at a vertex can also have an important bearing on the outcome of the test. The basic idea is that the domain be covered by a collection of overlapping patches, small enough that the solution of the problem can be approximated sufficiently well by a polynomial of degree m on each patch, the patch test should then provide an adequate test of the chosen interpolant. For a discussion of patches and features relating to their definition, we refer to Carey (1976).

One should bear in mind, when applying the patch test, that it originated in structural mechanics and, though satisfactory for the equations met in this context, it may require some amendments or supplementary conditions, in other situations. (See for example the remarks of Irons, 1983, in connection with the examples of Stummel, 1980, which we describe in section 4.) In practical situations, the role of the patch test is as a prerequisite for a suitable method; following a satisfactory outcome, the element is deemed to be capable of responding as a real material on the given mesh. Further tests are then invoked before the element is finally validated. That is, the patch test is a necessary but not a sufficient condition for acceptability. This contrasts with the mathematical view that it is not a necessary condition for convergence but it may be sufficient. We will return to this and related issues in Sections 4 and 5. For equations with variable coefficients, the test is normally conducted with the coefficients 'frozen' on each patch.

As we have mentioned previously, the patch test for the Poisson problem may be checked for a given element and specific grid by choosing any $u \in P_1$, determining f so that the differential equation is satisfied ($f \equiv 0$ in this instance) and then solving the discrete system

(2.8). If it is found that $U \equiv u$ for all patches, then the test is passed. Perhaps a more satisfactory approach would be to analytically evaluate the difference $a_*(V,u) - (V,f)$, with the respective integrals defined over patches, for all $u \in P_1$. This difference will be identically zero when V is any conforming function and it is therefore only necessary to consider strictly nonconforming functions. Since $\partial u/\partial \nu$ is constant on each edge for any $u \in P_1$, we conclude from (2.12) and (2.13), that satisfaction of the patch test is ensured if either

$$\int_{\partial K} \frac{\partial u}{\partial \nu} V \, ds = 0 \tag{2.14}$$

for each element, or

$$\int_{edge} [V] \, ds = 0. \tag{2.15}$$

for each edge. (The local nature of the basis functions in the finite element method does not generally allow a more elaborate means of cancellation of terms in (2.12) and (2.13).)

We now turn to the consideration of the biharmonic problem (2.2). Though its treatment broadly follows that given above, there are one or two technical difficulties which arise. It appears natural to define the nonconforming method in terms of the bilinear form

$$a_*(v,u) = \sum_{K\in\Omega} \int_K \nabla^2 v \, \nabla^2 u \, dK.$$

However, this form is not in general S^h- elliptic (see Ciarlet, 1978) and it is more convenient to define

$$a_*(v,u) = \sum_{K\in\Omega} \int_K \left[\nabla^2 v \, \nabla^2 u + (1-\sigma)\left[2v_{xy}u_{xy} - v_{yy}u_{xx} - v_{xx}u_{yy}\right]\right] dx\,dy \tag{2.16}$$

where $u_{xy} \equiv \partial^2 u/\partial x \partial y$, etc., and $\sigma \in [0,1/2)$ is Poisson's ratio. The additional terms in (2.16) vanish for $u,v \in H_0^2(\Omega)$ and can therefore be included without affecting the corresponding differential equation (2.2a). (Such a change will of course affect the natural boundary conditions of the problem but this does not concern us here.) To check the patch test, we take any $u \in P_2$, assign f so as to satisfy (2.2a) (this again implies $f \equiv 0$) and we find, by the application of Green's Theorem

$$a_*(V,u) - (V,f) = \sum_{K\in\Omega} \int_{\partial K} \left[(1-\sigma)\left[u_{\nu t}V_t - u_{tt}V_\nu\right] - \nabla^2 u_\nu V + \nabla^2 u \, V_\nu\right] ds \tag{2.17}$$

where u_t and u_ν denote, respectively, the derivatives of u in the tangential and normal directions to ∂K. We shall make use of this expression in the next section.

3. EXAMPLES OF NON-CONFORMING ELEMENTS

The finite element literature abounds with a wide variety of non-conforming elements of which it will be possible to describe only a representative sample (see Zienkiewicz, 1977, for further examples).

a) The non-conforming linear element

On a typical triangular element, the local interpolant is a linear function expressed in terms of its values at the mid points of each side (see Fig. 1a). The global form of the interpolant is therefore nonconforming for second order problems, being only midside continuous. Satisfaction of the patch test follows easily once it is observed that, for any $V \in S^h$, the jump [V] across an edge is a linear function of the tangential coordinate which vanishes at the midpoint of the edge. Consequently

$$\int_{\text{edge}} [V]\, ds = 0$$

and the patch test is passed. The convergence rates of this element are those normally expected of linear elements, $O(h)$ in energy and $O(h^2)$ in L_2.

Fig. 1 a) The nonconforming linear element and b) Wilson's element

b) Wilson's element

To the continuous bilinear interpolant, denoted by U_{bil}, are added two nonconforming functions on each element. For the $2h\times 2h$ element $\{|x-\bar{x}|\le h,\ |y-\bar{y}|\le h\}$ with centre at $(\bar{x},\bar{y})$, define $\phi(x) = 1-(x-\bar{x})^2/h^2$. The interpolant U for Wilson's element then has the form

$$U = U_{bil} + \alpha\phi(x) + \beta\phi(y) \qquad (x,y) \in K \tag{3.1}$$

where we may interpret the parameters α and β as the average values of the derivatives u_{xx} and u_{yy}, respectively, over the element. Note that the global interpolant is continuous at the vertices of each element. With $V = \phi(y)$ for example, we find

$$\int_{\partial K} \frac{\partial u}{\partial \nu} V\, ds = \int_{\bar{y}-h}^{\bar{y}+h} \frac{\partial u}{\partial x} \phi \Big|_{\bar{x}+h} dy - \int_{\bar{y}-h}^{\bar{y}+h} \frac{\partial u}{\partial x} \phi \Big|_{\bar{x}-h} dy. \tag{3.2}$$

This expression vanishes when u is a linear function since the integrands take the same values on the vertical edges $x = \bar{x} \pm h$ of the element. The condition (2.14) is therefore satisfied and the patch test passed. For the cancellation of terms in (3.2) to occur the rectangular shape of the element is crucial and no distortions, other than to parallelogram shapes, are permissible. We shall see in Section 5 that, by using Stummel's generalised patch test, convergence is possible under somewhat weaker conditions.

One possible reason for introducing nonconforming terms can be seen by examining the gradient of U. From (3.1)

$$\underline{\nabla U} = \begin{bmatrix} b+dy \\ c+dx \end{bmatrix} - \frac{2}{h^2} \begin{bmatrix} \alpha(x-\bar{x}) \\ \beta(y-\bar{y}) \end{bmatrix} \tag{3.3}$$

when $U_{bil} = a + bx + cy + dxy$. Each component of the gradient of the conforming part of the solution contains linear functions of only one variable. Since the finite element method obtains the best L_2 fit to u in gradient, the form (3.3) is expected to be more satisfactory for this purpose. For a thorough analysis of this element in three space dimensions we refer to Ciarlet (1978).

An analogous element, applicable to triangles, may be constructed by adding nonconforming functions to the usual continuous piecewise linear approximation. On each element, these nonconforming functions take the form

$$\lambda_1\lambda_2 + \lambda_2\lambda_3 + \lambda_3\lambda_1 \tag{3.4}$$

where λ_1, λ_2 and λ_3 are the barycentric or area coordinates for the triangle. These functions have disjoint support and each is associated with a node located at the centroid of the element. The patch test is passed for each function V of the form (3.4) by virtue of the fact that

$$\int_{\partial K} V \frac{\partial u}{\partial \nu} ds = 0 \qquad \forall u \in P_1.$$

(See Carey, 1976). For Poisson problems, the conforming and non-conforming contributions are orthogonal to each other and can therefore be computed separately.

The three examples we have presented so far have each illustrated a different mechanism for satisfying the patch test. In the first, the nonconforming contributions cancelled across each edge, in the second they cancelled across opposite edges and in the third example, the sum of contributions along all edges cancelled.

c) The Morley element

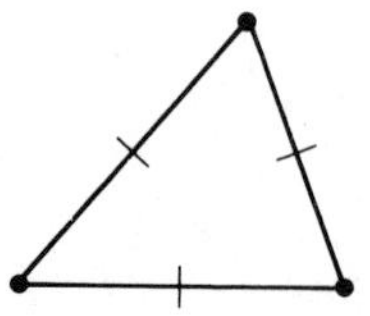

Fig. 2 The Morley element

This is one of the simplest elements available for the approximation of fourth order biharmonic type problems. The local interpolant U is a quadratic function expressed in terms of its values at the vertices and normal derivatives at the mid-side points. Though the global interpolant is not even continuous across element boundaries (except at the vertices) it nevertheless leads to convergent approximations to both second and fourth order problems.

To check that the patch test is satisfied for our fourth order example we are led, by (2.17), to consider the expression

$$\int_{edge} \left\{(1-\sigma)\{u_{tt}[v_\nu] + u_{\nu t}[v_t]\} - \nabla^2 u_\nu[v] + \nabla^2 u[v_\nu]\right\} ds, \quad (3.5)$$

which represents the contribution made to (2.17) from each edge. For $u \in P_2$, the derivatives of u appearing in (3.5) are each constant and

$$\int_{edge} [v_\nu]\, ds = \int_{edge} [v_t]\, ds = 0$$

which result from the fact that the integrands are linear functions of the tangential coordinate that vanish at the midpoint of each edge. The details of the calculation may be verified by making use of the explicit forms of the local interpolants on two adjacent elements given in Mitchell and Wait (1977).

In slow flow problems the velocity field q and the pressure p satisfy the Stokes equation

$$\nabla^2 q = \nabla p + f$$

and the continuity equation

$$\nabla \cdot q = 0$$

Introducing a stream-function ψ by $q = \text{curl}\ \psi$ it is easily shown that ψ satisfies the biharmonic equation. Approximation of ψ by the Morley element is equivalent to a mixed approximation of the Stokes equations using the nonconforming linear element (Example a) for the velocity and piecewise constant representation of the pressure (Crouzieux and Raviart, 1973, Temam, 1977, Griffiths, 1979). The design of non-conforming elements for high order problems can therefore often be aided by considering a related problem of lower order. This process is not without its difficulties since one must meet the so-called Babuska-Brezzi condition for the compatibility of the velocity and pressure spaces (see Chapter 7).

d) A nonconforming bilinear element

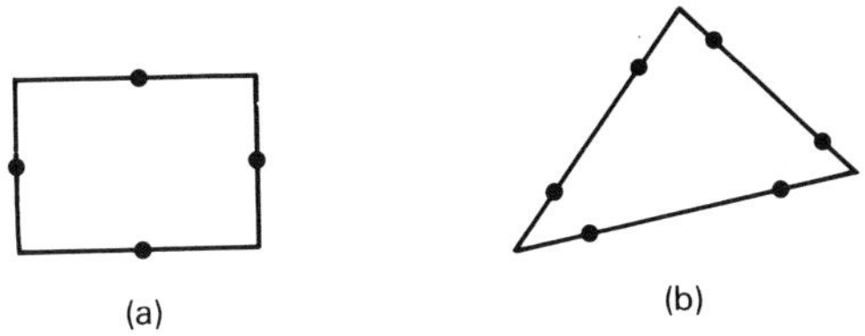

Fig. 3 Nonconforming elements of a) bilinear and b) quadratic types

This element, applicable to second order problems, illustrates some new kinds of difficulties associated with nonconforming elements. With nodes placed at the midside points (Fig. 3a) it is not possible to define a unique bilinear interpolant on this element directly because there is a function of the form $(x-a)(y-b)$ which vanishes at all four nodes. Fortin and Soulie (1983) suggested how this could be overcome by writing the local interpolant in the form

$$U = U_{bil} + \alpha(x-\bar{x})(y-\bar{y}) \tag{3.6}$$

where U_{bil} is again the usual continuous bilinear interpolant, α a parameter and $(\bar{x},\bar{y})$ the coordinates of the centroid. Clearly U is a linear function along each edge and is midside continuous. This example consequently passes the patch test in the same way as example a). The representation (3.6) may be used to compute the approximation U provided that Dirichlet boundary conditions are imposed by interpolation at the midside nodes on $\partial\Omega$. If, on a square grid of size h, we denote by W_{ij} the nodal values of U_{bil} at (ih,jh), then it may be shown that

$$4\,W_{ij} - W_{i+1,j+1} - W_{i+1,j-1} - W_{i-1,j+1} - W_{i-1,j-1} = 2h^2 \tag{3.7}$$

and, on the element $\{ih \le x \le ih+h;\ jh \le y \le jh+h\}$, α is given by

$$-W_{ij} + W_{i+1,j} - W_{i+1,j+1} + W_{i,j+1}$$

for the special case of the Poisson problem $-\nabla^2 u = 1$. The interior equations therefore split into separate finite difference schemes, one on each of the 'red' and 'black' squares in a checkerboard pattern. The two sets of equations are, however, coupled at the boundary. For instance, at $i = 0$, the boundary condition reads

$$W_{0,j+1} + W_{0,j} = 2u(0,(j+1/2)h). \tag{3.8}$$

Closer inspection reveals that the interpolated Dirichlet conditions on $\partial\Omega$ are insufficient to specify W uniquely (there is one degree of freedom which may be removed by fixing W arbitrarily at any one boundary node) but when these (non-unique) values are used to construct U_{bil} and the result substituted into (3.6), the final expression for U is unique.

It is well-known (see for instance Ciarlet, 1978, p. 206) that the same difference equations (3.7) are obtained with a conforming bilinear approximation when the integrals that occur are evaluated by one point Gauss quadrature. There are of course differences in the way the two methods treat the boundary conditions.

e) A nonconforming quadratic element

Our final example, also studied by Fortin and Soulie (1983), takes as its starting point the standard 6-node conforming quadratic triangle and adds, on each element, a multiple of the quadratic function which vanishes at the two Gauss points of each side. The resulting element may be represented by the nodal configuration shown in Fig. 3b. The development of this element broadly follows that of the previous example. It has, in addition, a number of fascinating properties that are described by Fortin and Soulie who also apply the element to the equations of both potential and viscous flow.

4. A COUNTEREXAMPLE TO THE PATCH TEST?

Since its inception in 1965 there has been considerable speculation in both engineering and mathematical communities concerning the relationship between the patch test and convergence. It is perhaps fair to say that until recently, the commonly held view, as expressed for instance by Strang and Fix (1973), was that the patch test is sufficient though not necessary for convergence. This position has had to be reviewed following the paper of Stummel (1980) which contained an example which purported to pass the patch test and yet it converged to the wrong solution. Before commenting further on Stummel's results, it is necessary to describe the example in some detail.

We consider therefore the one dimensional boundary value problem with, say, $f \in L_2(0,1)$

$$-u'' + u = f \quad \text{for} \quad 0 < x < 1 \tag{4.1}$$

with the Dirichlet boundary conditions

$$u(0) = g_0, \qquad u(1) = g_1. \tag{4.2}$$

Defining $g(x) = (1-x)g_0 + xg_1$, the weak form of the problem can be specified by:- find $u-g \in H_0^1(0,1)$ such that

$$a(v,u) = (v,f) \qquad \forall v \in H_0^1 \tag{4.3}$$

where the bilinear form

$$a(v,u) = \int_0^1 \{v' u' + v u\} dx \tag{4.4}$$

is continuous and H_0^1 - elliptic.

We now divide the interval (0,1) into N equal elements of length $h = 1/N$ and introduce two spaces of piecewise polynomials. The first of these is a conforming space S_C^h of continuous piecewise linear functions which vanish at $x = 0,1$ and the second is a nonconforming space S_N^h of piecewise constant functions which vanish outside the interval $(h,1-h)$. The full trial space is then the direct sum: $S_0^h = S_C^h \oplus S_N^h$. We define the modified bilinear form on $S_0^h \times S_0^h$ in the usual way:

$$a_*(V,U) = \sum_{j=1}^{N-1} \int_{jh}^{jh+h} \{V' U' + V U\} dx. \tag{4.5}$$

It is readily verified that $|U|_* = \{a_*(U,U)\}^{1/2}$ is a norm on S_0^h so that the bilinear form (4.5) is automatically S_0^h - elliptic. The nonconforming approximation to (4.3) is defined by:- find $U-g \in S_0^h$ such that

$$a_*(V,U) = (V,f) \qquad \forall V \in S_0^h. \tag{4.6}$$

It is convenient to isolate the conforming and nonconforming parts of U by writing $U = g + U_C + U_N$ with $U_C \in S_C^h$ and $U_N \in S_N^h$. The system (4.6) may now be written in as a pair of coupled problems:
Find U_N and U_C such that

$$(V', U_C'+g') + (V,U) = (V,f) \qquad \forall V \in S_C^h \tag{4.7}$$

and

$$(W,U) = (W,f) \qquad \forall W \in S_N^h. \tag{4.8}$$

Choosing $V \equiv U_C$ and $W \equiv U_N$ in (4.7) and (4.8) and adding the result gives

$$\|U_C'\|^2 + (U,U-g) = (U-g,f) \tag{4.9}$$

where $\|\cdot\|$ denotes the L_2 norm and we have used the fact that $(V',g') = 0$. Rearranging, we find

$$\|U_C'\|^2 + \|U - \frac{f+g}{2}\|^2 = \frac{1}{4}\|f-g\|^2. \tag{4.10}$$

The patch test can be verified directly from this identity. If $u = p_1 \in P_1$ is a solution of (4.1) then $f = g = p_1$. Thus the right side of (4.10) vanishes giving

$$U_C' \equiv 0, \qquad U = \frac{1}{2}(f+g) = p_1$$

so that $U = u$ as required for satisfaction of the patch test.

To study convergence under more general conditions, let F denote an arbitrary element of S_N^h. By virtue of (4.8) we have $(F,U-f)=0$ and, adding this quantity to the left of (4.9), gives

$$\begin{aligned}\|U_C'\|^2 + \|U-f\|^2 &= (g-f-F,U-f)\\ &\le \frac{1}{2}\|g-f-F\|^2 + \frac{1}{2}\|U-f\|^2\end{aligned}$$

and hence

$$\|U_C'\|^2 + \frac{1}{2}\|U-f\|^2 \le \frac{1}{2}\inf_{F\in S_N^h}\|g-f-F\|^2. \tag{4.11}$$

The right hand side of (4.11) represents the error in the best L_2 approximate to $(g-f)$ by piecewise constant functions and thus tends to zero with h. Hence $U_C \to 0$ and $U \to f$ as $h \to 0$ which is manifestly not the solution to (4.1).

Some remarks are appropriate at this stage.

i) The inclusion of just one additional function to S_0^h, for example,

$$\begin{cases} x - h/2, & 0 < x < h \\ 0, & h < x < 1 \end{cases},$$

would make it the most general piecewise linear function on the grid and it would not then satisfy the patch test.

ii) It was not strictly accurate, as we claimed, that this element passes the patch test for it was demonstrated only when the patch constituted the entire domain. Any attempt to take a patch consisting of fewer elements would fail due to a lack of uniqueness brought about because there are two ways of constructing a constant function on a patch. This was recognised by Stummel (1980) who then modified the nonconforming functions to try and accommodate this fact. We shall describe a slightly different modification which displays the same convergence properties. In fact we replace the nonconforming space S_N^h by one which spans the piecewise constant functions with restricted support:-

$$\begin{cases} 1, & (j+\lambda)h < x < (j+1-\lambda)h \\ 0, & \text{otherwise} \end{cases} \qquad j = 0,1,..,N-1 \tag{4.12}$$

where $0 < \lambda < 1/2$. By arguments similar to those given above it may be proved that the conforming part of the approximation converges in L_2 to the solution, w, of the boundary value problem

$$-w'' + 2\lambda(w-f) = 0, \qquad w(0) = w(1) = 0$$

whilst the overall approximation U converges weakly in L_2 to $g + 2(1-\lambda)w$. (See Stummel (1980) for the case $\lambda = 1/3$ and Shi (1983) for $\lambda = m/(2m+1)$, m integer.) Thus convergence always takes place to the wrong solution unless $\lambda = 1/2$, i.e. the nonconforming functions vanish almost everywhere.

We have, in this discussion, overlooked the fact that the bilinear form (4.5) is not defined for functions such as (4.12) unless the ranges of the integrals are further broken down to avoid the new discontinuities which have been introduced into U. This brings us into the realm of macro-elements where the usual convention (see for instance Ciarlet, 1978) is that the basis functions possess at least as much smoothness inside each element as they do at element boundaries. This brings into question whether or not Stummel's examples constitute elements as they are commonly understood.

iii) In the two examples we have presented in this section, the finite element approximation of the term -u" has two components: (V', U') in S_C^h and zero in S_N^h. Both are acceptable insofar as the patch test is concerned for then u is a linear function and the required value zero is obtained in both cases (note that the presence of the term u in (4.1) is essential to maintain stability). However, when u is no longer linear, the approximation to -u" from S_N^h is patently useless: the patch test has therefore failed to test the correct properties of the scheme. In fairness, we should point out that the definition of the patch test given in Section 2 is the mathematical interpretation given by Strang and Fix (1973) to the more physically based test proposed by Irons. Furthermore, Strang and Fix, were primarily concerned with Laplace and Biharmonic type problems where the differential operators involved only the highest order terms. Were we, in the present context, to apply the Stummel spaces to the problem

$$-u'' = f, \quad u(0) = g_0, \quad u(1) = g_1$$

the patch test would fail because the appropriate bilinear form would vanish identically on $S_N^h \times S_N^h$ and the discrete equations on any patch would not have unique solutions. Since non-conforming elements are, by definition, those which fail to meet the continuity requirements imposed by the highest derivatives appearing in the problem, it is perhaps natural therefore that the patch test, in the first instance, should be confined only to those terms. We believe, that in this way, faith in the patch test may be restored. In a recent paper, Irons (1983) has responded to Stummel's examples by stating the patch test, and the use to which the results are put by engineers, more precisely. Though these are again phrased in physical terms, we feel that they are in broad agreement with our remarks in this paragraph.

iv) To remove any impression that discontinuous elements may not be allowed in one space dimension, we introduce one further example in which the conforming space is as previously defined and the nonconforming space, S_N^h, is spanned by the piecewise quadratic functions

$$\psi_j(x) = \begin{cases} (2x/h-2j-1)^2 - 1/3, & jh < x < jh+h \\ 0, & \text{otherwise} \end{cases} \tag{4.13}$$

for $j = 0,1,..,N-1$. With this basis, S_N^h is actually orthogonal to S_C^h with respect to (4.5) and thus the computation of U breaks down into two separate parts:- find $U_C \in S_C^h$ such that

$$a_*(V,U_C) = (V,f) \quad \forall V \in S_C^h$$

which gives the usual piecewise linear approximation to (4.1). We also have:- find $U_N \in S_N^h$ such that

$$a_*(W,U_N) = (W,f) \quad \forall W \in S_N^h$$

which can be computed element by element. In fact, if $W = \psi_j$ it is readily shown that $(W,f) \simeq Ch^3 f''(\xi)$, $jh < \xi < jh+h$, so that

U_N provides an $O(h^2)$ correction to U_C and the jumps in U at the knots have a magnitude $O(h^2)$.

To conclude this section we describe a link between the first example of this section and mixed finite elements. In particular, the solutions $(U_C, U_N) \in S_C^h \times S_N^h$ of equations (4.7) and (4.8) provide a mixed conforming approximation to the problem of finding $(v,w) \in H_0^1 \times L_2$ which satisfy the weak forms of the equations (see Chapter 7)

$$-v'' + v + w = f - g, \qquad v(0) = v(1) = 0$$

$$v + w = f - g$$

whose solution is clearly $w = f-g$ and $v = 0$, i.e. the values to which U_N and U_C were previously shown to converge.

5. THE GENERALISED PATCH TEST

From a mathematical point of view, the position of the Irons - Strang patch test is rather ambivalent. It has been known for some time not to give necessary conditions for convergence and, by the examples of the previous section, there is some doubt whether or not it is sufficient (we cannot be more definite in this case because it depends on whether or not the approximation is classified as being a finite element method). This version of the test, therefore, plays no part in proofs of convergence of nonconforming methods and these have to proceed independently in each case by estimating the quantity $\Delta(u)$ defined by (2.10). Recently, Stummel (1979) has introduced a generalised form of the patch test which, subject to a certain approximation property, provides both necessary and sufficient conditions for convergence. In contrast to the original test, it does not decide on the suitability of the trial space for a given differential operator, rather, it tests compatibilty of the trial space with the underlying function space in which the problem is posed. For example, if a given trial space S^h is compatible with the Sobolev space $H^m(\Omega)$ in the limit $h \to 0$, then, subject to the approximation property being satisfied, S^h will give rise to convergent approximations to all elliptic problems of order 2m with natural boundary conditions (when the space is $H_0^m(\Omega)$, the associated boundary conditions are of Dirichlet type). The greater precision and wider applicability of the results are, however, gained at the expense of having a test that is more difficult to apply.

To illustrate Stummel's results, we consider the 2mth order elliptic equation

$$\sum_{|\sigma|,|\tau|\leq m} (-1)^{|\sigma|} D^\sigma(a_{\sigma\tau} D^\tau u) = f \quad \text{in } \Omega \tag{5.1}$$

where Ω is an open polyhedral domain in $\mathbb{R}^n$, $f \in H^{-m}(\Omega)$ (the dual space of H^m) and the coefficients $a_{\sigma\tau}$ are bounded measurable functions on Ω.

We have made use of the usual multi-index notation where σ, τ are n - arrays of non-negative integers, for example, $\sigma = (\sigma_1, \sigma_2, .., \sigma_n)$ and $|\sigma| = \Sigma\, \sigma_i$. Furthermore,

$$D^{\sigma} u \equiv \frac{\partial^{|\sigma|} u}{\partial^{\sigma_1} x_1 \partial^{\sigma_2} x_2 .. \partial^{\sigma_n} x_n}$$

denotes the generalised derivative of order $|\sigma|$.

We shall suppose, for simplicity, that (5.1) is subject to homogeneous Dirichlet boundary conditions on u and its first m-1 normal derivatives. The weak form of (5.1) then amounts to solving for $u \in H_0^m$ from

$$\sum_{|\sigma|,|\tau| \leq m} \int_{\Omega} a_{\sigma\tau} D^{\sigma} v\, D^{\tau} u\, d\Omega = \int_{\Omega} v\, f\, d\Omega \quad \forall v \in H_0^m. \tag{5.2}$$

For each h in a sequence tending to zero, assume that we have a decomposition of Ω into regular elements (in the usual finite element sense) having a notional diameter h and a space of piecewise polynomial functions $V \in S_0^h$. We define the semi-norms

$$|V|_{r,K} \equiv \left\{ \sum_{|\sigma|=r} \int_K |D^{\sigma} V|^2\, d\Omega \right\}^{1/2},$$

$$|V|_{r,\Omega} \equiv \left\{ \sum_{K \in \Omega} |V|_{r,K}^2 \right\}^{1/2}$$

and the associated norms,

$$\|V\|_{m,K} \equiv \left\{ \sum_{r \leq m} |V|_{r,K}^2 \right\}^{1/2}$$

$$\|V\|_{m,\Omega} \equiv \left\{ \sum_{r \leq m} |V|_{r,\Omega}^2 \right\}^{1/2}.$$

(The norm $|\cdot|_{1,\Omega}$ is the same as $|\cdot|_*$ introduced in Section 2.) The approximation property that we alluded to earlier can now be stated more precisely. It requires that there exist a natural number $r > m$ and a constant C, independent of h, such that to each $u \in H^r(\Omega)$ we have functions $V \in S^h$ satisfying the error estimate

$$\|u - V\|_{m,K} \le C\, h^{r-m} |u|_{r,K} \tag{5.3}$$

uniformly for all $K \in \Omega$. This condition will normally be met if the local interpolants contain complete polynomials of degree $(r-1)$.

In order to satisfy Stummel's generalised form of the patch test, we must have, for bounded sequences of functions $V \in S_0^h$,

$$\lim_{h \to 0} \sum_{K \in \Omega} \int_{\partial K} \psi \, D^{\mu} V \, \nu_k ds = 0 \tag{5.4}$$

for all test functions $\psi \in C_0^{\infty}(\mathbb{R}^n)$ and all indices $k = 1,2,..n$ and $|\mu| \le m-1$ (ν_k denotes the kth component of the unit outward normal to ∂K). Slight modifications are necessary when different boundary conditions are imposed and we refer to Stummel (1979) for details.

We now re-examine some of the earlier examples in the light of this test. For the one-dimensional problem (4.1), $m = 1$ and (5.4) reduces to showing that the expression

$$\sum_{j=1}^{N} \psi(x_j) V(x) \Big|_{x_{j-1}+0}^{x_j-0} \tag{5.5}$$

has limit zero as $h \to 0$ for all $\psi \in C_0^{\infty}(\mathbb{R})$ and for all sequences $V \in S_0^h$ bounded with respect to the norm $|\cdot|_{1,\Omega}$. Clearly the conforming part of the approximations used in Section 4 gives no contribution to (5.5) and it is, as usual, only the nonconforming part that need concern us further. For the first example of Section 4 the nonconforming functions V were constant on each element and the expression (5.5) may be re-written in the form

$$\int_0^1 V \psi' \, dx \tag{5.6}$$

and cannot have the correct limiting behaviour. The test is therefore failed and this approximation would not be expected to converge.

It is instructive to compare the above procedure with a formal application of the patch test. With $u = a + bx \in P_1$ and $f = -u'' + u$ we find, for the patch $0 < x < 1$,

$$a_*(V,u) - (v,f) = \sum_{j=1}^{N} \int_{jh-h}^{jh} \{V' u' + Vu\} dx - \int_0^1 V\{-u'' + u\} dx \tag{5.7}$$

$$= \sum_{j=1}^{N} b \int_{jh-h}^{jh} V' \, dx. \tag{5.8}$$

This expression vanishes for any piecewise constant function V; it also corresponds to (5.5) in the special case $\psi(x) \equiv b$ ($\psi(x)$ can easily be modified outside (0,1) so that it has compact support on the real line). To estimate the quantity $\Delta(u)$ given by (2.10), we would need to evaluate (5.7) for more general functions u. In these circumstances (5.8) would be replaced by

$$\sum_{j=1}^{N} \int_{jh-h}^{jh} u' V' \, dx \tag{5.9}$$

from which it would follow that $\Delta(u) \to 0$ as $h \to 0$ if the generalised patch test were passed (this is, of course, not the case here).

We turn now to the two dimensional example provided by Wilson's element (Example b of Section 3). For second order problems, we take $\mu = 0$ in (5.4) and, as before, the test is automatically fulfilled for the conforming part of the approximation. We need therefore only consider local approximants of the form (see 3.1)

$$V = \alpha \, \phi(x) + \beta \, \phi(y), \qquad (x,y) \in K.$$

For the vertical edges of an element, $\nu = (\pm 1, 0)$ and, with $k = 1$, the contribution to (5.4) from a typical element $K=\{(x,y); |x-\xi|\le h, \ |y-\eta|\le h\}$ is

$$\int_{\partial K} \psi \, V \, \nu_1 \, ds = \int_{\eta-h}^{\eta+h} \psi V \Big|_{\xi-h}^{\xi+h} dy. \tag{5.10}$$

However, V has identical values on these edges and, moreover,

$$\psi(\xi+h,y) - \psi(\xi-h,y) = \int_{\xi-h}^{\xi+h} \frac{\partial\psi}{\partial x} \, dx.$$

Hence the right hand side of (5.10) becomes (cf. equation (3.2))

$$\int_K V \frac{\partial\psi}{\partial x} \, dxdy. \tag{5.11}$$

Using the Cauchy-Schwartz inequality we have

$$\begin{aligned} \left| \int_K V \frac{\partial\psi}{\partial x} \, dxdy \right| &\le \|V\|_{0,K} \|\psi_x\|_{0,K} \\ &\le \|V\|_{0,K} |\psi|_{1,K}. \end{aligned} \tag{5.12}$$

We note the close similarity between (5.11) and the analogous expression (5.6) for the one dimensional problem. However, the results are more positive in the present case because of the increased dimensionality: since V vanishes on the horizontal edges of the grid, we may apply a Poincaré type inequality (see Strang and Fix, 1973) to give

$$\|V\|_{0,K} \leq Ch|V|_{1,K}.$$

Substituting this into (5.12) and summing the result over all $K \in \Omega$, we obtain

$$\left| \sum_{K\in\Omega} \int_K V \frac{\partial\psi}{\partial x} \, dxdy \right| \leq Ch|V|_{1,\Omega}|\psi|_{1,\Omega}, \tag{5.13}$$

where the right hand side tends to zero with h for all $\psi \in C_0^\infty(\mathbb{R}^2)$ and all sequences $V \in S_0^h$ that are bounded in the sense that $|V|_{1,\Omega} \leq C < \infty$. The analysis is entirely similar in the case $k = 2$ and the test is therefore passed and convergence assured.

The analysis of the Wilson element for quadrilateral shapes than parallelograms is considerably more difficult. Convergence has been proved by Shi (1983a) provided that the shapes tend to parallelograms as $h \to 0$. This is the case, for instance, when the sequence of subdivisions is constructed by dividing each element into four by joining the midpoints of opposite sides. Shi has also proved that this qualification is in some sense necessary for convergence. The achievement of satisfactory accuracy in a practical calculation is conditional therefore on the grid being composed of elements that are only slight distortions of parallelograms. This may be particularly difficult to achieve without an extremely fine grid if the geometry of the region is complex.

In a recent series of papers, Shi (1983a - e) has obtained similar results for a variety of elements which fail to meet the Irons - Strang version of the patch test. As in the Wilson element, in the limit $h \to 0$ the element shapes have to be parallelograms and, interestingly, the original patch test is satisfied in such instances, i.e. on infinitesimal patches. The elements analysed by Shi include those proposed and studied numerically by Sander and Beckers (1977). In each case the numerically observed convergence or divergence of an element is predictable according to Stummel's generalised form of the patch test.

6. NON-SELFADJOINT PROBLEMS

The increasing interest in the approximation of non-selfadjoint problems has been brought about largely because of their importance in fluid dynamics. We shall illustrate some of the variations of the finite element method that are in common use in this area by means of the convection - diffusion equation

$$-\varepsilon\nabla^2 u + b\cdot\nabla u = f \quad \text{in } \Omega \tag{6.1}$$

where $\varepsilon > 0$ and $b \in \mathbb{R}^n$ ($n = 1$ or 2) are constant and, since we are principally concerned with the effects of nonconformity, restrict ourselves to the case of homogeneous boundary conditions

$$u = 0 \quad \text{on } \partial\Omega \tag{6.2}$$

though it is of little interest in applications. The inhomogeneous case of (6.2) may be treated by the same device that was used in Section 2.

Standard Galerkin methods, whether conforming or otherwise, are perfectly satisfactory from the point of view of convergence. Difficulties arise in practice because, in order to meet the theoretical requirements, the mesh must be unrealistically fine. Typically the grid size h should not exceed some fixed multiple of $\varepsilon/|b|$. When h exceeds this threshold the numerical solution develops spurious oscillations and is of little or no value. In order to develop viable techniques for small values of ε, the finite element method has been generalised by allowing different test and trial functions and has become known as the Petrov - Galerkin method. An elegant framework for determining optimal combinations of test and trial functions is provided by Morton in Chapter 6 of this volume and we shall therefore concentrate in the main on issues relating to nonconformity.

Retaining the notation introduced in Section 2, the weak form of (6.1) and (6.2) is:- find $u \in H_0^1(\Omega)$ such that

$$a(v,u) = (v,f) \quad \forall v \in H_0^1(\Omega) \tag{6.3}$$

where $a(\cdot,\cdot)$ is the unsymmetric bilinear form defined by

$$a(v,u) = \int_\Omega \{\varepsilon\nabla v\cdot\nabla u + v(b\cdot\nabla u)\}\, d\Omega. \tag{6.4}$$

To approximate this problem we introduce a trial space S_0^h, which will contain the finite element solution U, and a test space T_0^h with which to approximate the test functions v in (6.3). The idea behind this device, introduced into convection - diffusion problems by Christie et al (1976), is to try to choose the test space, for a specific choice of trial space, so as to counteract the asymmetry present in (6.4). With $S_0^H \subset H_0^1(\Omega)$ and $T_0^h \subset H_0^1(\Omega)$, we define the conforming Petrov - Galerkin approximation of (6.3) to be:- find $U \in S_0^h$ such that

$$a(V,U) = (v,f) \quad \forall V \in T_0^h \tag{6.5}$$

and a necessary condition that this system have a unique solution is that the test and trial spaces have equal dimension. Whilst it is certainly possible for both test and trial spaces to be nonconforming by having jump discontinuities at element boundaries, this generally allows more flexibility than can usefully be exploited. It is sufficient in most cases to choose the trial space to be one of the classical, conforming, low order finite elements of Lagrange type. The local approximants for the test space are then constructed by perturbing those of S_0^h either by continuous high order piecewise polynomials (leading to a conforming method) or by discontinuous low order piecewise polynomials (the discontinuities will be assumed to be confined to inter-element boundaries). There is little to choose between these two approaches on the grounds of accuracy, in fact it can often be arranged that the resulting algebraic equations in the two cases are the same. With lower order discontinuous test functions the integrals which occur during the course of the calculation are cheaper to evaluate and, as we shall show presently, the relationship between T_0^h and S_0^h is easier to define. If we are to choose $S_0^h \subset C^0(\Omega)$ and $T_0^h \not\subset C^0(\Omega)$, is $a(\cdot,\cdot)$ defined on $T_0^h \times S_0^h$? Referring to (6.4) we see that the integrand will involve products of δ-functions and Heaviside functions (the Heaviside function $H(x)$ is defined to have the value 1 for $x>0$ and zero for $x<0$). The integrals of such products are not defined in the usual sense but they can be attributed finite non-zero values within the context of generalised functions by making use of the definitions of Jones (1980, 1982). (Recall that nonconforming Galerkin finite elements led to integrals of the squares of δ-functions; these can also be interpreted in this way and give a zero contribution.) Therefore, defining

$$a_*(V,U) = \sum_{K\in\Omega} \int_K \{\varepsilon\nabla V\cdot\nabla U + V(b\cdot\nabla U)\}\, dK \tag{6.6}$$

we have that $a_*(V,U) \neq a(V,U)$ for $(V,U) \in T_0^h \times S_0^h$, since (6.6) ignores the contributions from inter-element boundaries. In these situations we choose to replace (6.5) by the discrete system

$$U \in S_0^h: \quad a_*(V,U) = (V,f) \quad \forall V \in T_0^h \tag{6.7}$$

if only because it is more convenient to deal with.

Existence and uniqueness of solutions of (6.7) follow from the generalised Lax-Milgram Theorem (Babuska and Aziz, 1972) if

i) (Continuity)

$$|a_*(V,U)| \le C_1 |V|_{1,\Omega} |U|_{1,\Omega}, \qquad C_1 < \infty \tag{6.8}$$

ii) (Positivity)

$$\sup_{V \in T_0^h} \frac{|a_*(V,U)|}{|V|_{1,\Omega}} \geq C_2|U|_{1,\Omega}, \quad C_2 > 0 \tag{6.9}$$

iii)

$$\sup_{U \in S_0^h} |a_*(V,U)| > 0. \tag{6.10}$$

With these assumptions, the analysis leading up to (2.11) for the Galerkin case can be repeated here and we find, defining

$$\Delta(u) = \max_{V \in T_0^h} \frac{|a_*(V,u) - (V,f)|}{|V|_{1,\Omega}} \tag{6.11}$$

that

$$C_1^{-1}\Delta(u) \leq |u - U|_{1,\Omega} \leq \left\{1 + C_1/C_2\right\} \inf_{W \in S_0^h} |u - W|_{1,\Omega} + C_2^{-1}\Delta(u) \tag{6.12}$$

showing that a necessary condition for convergence is that $\Delta(u) \to 0$ as $h \to 0$. It is interesting to note that this result does not depend on the approximating ability of the test space (the approximation properties of T_0^h are important in the derivation of L_2 estimates).

For Galerkin methods that satisfy the inverse assumption (Mitchell and Wait, 1976)

$$\|U\|_{L_2(\Omega)} \leq Ch\,|U|_{1,\Omega} \quad \forall U \in S_0^h \tag{6.13}$$

it is readily shown from (6.6) that the constants C_1, C_2 required by (6.12) satisfy (see equation (2.15) of Chapter 6)

$$C_1 \leq \varepsilon + Ch|b|, \quad C_2 \geq \varepsilon$$

and the principal contribution to the error comes from the fact that $C_1/C_2 (\leq 1 + Ch|b|/\varepsilon)$ is extremely large (the quantity $h|b|/\varepsilon$ is known as the mesh Peclet number). The method of artificial viscosity, analogous to upwinding in finite difference methods, tries to alleviate this problem by replacing ε by $\varepsilon + ch|b|$ in (6.6) for some positive constant c. Now the ratio

$$C_1/C_2 \le \frac{\varepsilon + Ch|b|}{\varepsilon + ch|b|} \qquad (6.14)$$

is much smaller than previously and, though the solutions are greatly improved, there is considerable smearing in the cross-flow direction caused by an excess of diffusion. The Streamline-Upwind method is one of the currently most effective ways of avoiding this smearing. When S_0^h is the space of continuous piecewise linear functions, test functions $V \in T_0^h$ are constructed so that

$$V = W + \gamma b \cdot \nabla W, \qquad W \in S_0^h \qquad (6.15)$$

and $\gamma > 0$ is a parameter proportional to the mesh spacing h. When S_0^h is based on bilinear elements, the test space is constructed by taking the tensor product of the one-dimensional elements (see Hughes and Brooks, 1979, Griffiths and Mitchell, 1979 and Chapter 6).

The test functions defined by (6.15) are clearly nonconforming; each is a piecewise constant perturbation of a trial function W. It is relatively easy to estimate the quantity $\Delta(u)$ in this case since $\nabla V = \nabla W$ on each element K. With $f = -\varepsilon\nabla^2 u + b\cdot\nabla u$ we have, by (6.15) and Green's Theorem

$$\begin{aligned} a_*(V,u) - (V,f) &= \sum_{K\in\Omega} \int_K \{\varepsilon\nabla W\cdot\nabla u + W(b\cdot\nabla u) + \gamma(b\cdot\nabla W)(b\cdot\nabla u)\}\, dK \\ &\quad - \int_\Omega \{Wf + \gamma(b\cdot\nabla W)f\}\, d\Omega \\ &\le \sum_{K\in\Omega} \int_K \gamma(b\cdot\nabla W)(b\cdot\nabla u)\, dK - \int_\Omega \gamma(b\cdot\nabla W)f\, d\Omega \\ &\le \gamma\varepsilon \int_\Omega \nabla^2 u(b\cdot\nabla W)\, d\Omega. \end{aligned} \qquad (6.16)$$

Hence,

$$|a_*(V,u) - (V,f)| \le \gamma\varepsilon|b|\,|u|_{2,\Omega}|W|_{1,\Omega}$$

and, from (6.11),

$$\Delta(u) \le \gamma\varepsilon|b|\,|u|_{2,\Omega}$$

since $|V|_{1,\Omega} = |W|_{1,\Omega}$. Thus $\Delta(u) \to 0$ as $h \to 0$ provided that $\gamma = O(h)$ and, moreover, the terms involving $\Delta(u)$ in (6.12) do not depend explicitly on ε. We note also that $\Delta(u) \equiv 0$ for $u \in P_1$ i.e. the patch test is passed. To determine the constants C_1 and C_2 for this method we find, by the Cauchy-Schwartz inequality and the inverse assumption (6.13), that

$$C_1 \le \varepsilon + \gamma|b|^2 + Ch|b|,$$

whilst

$$\sup_{V \in T_0^h} \frac{a_*(V,u)}{|V|_{1,\Omega}|U|_{1,\Omega}} = \sup_{W \in S_0^h} \frac{a_*(W+\gamma b\cdot\nabla W,U)}{|W|_{1,\Omega}|U|_{1,\Omega}}$$

$$\ge \frac{a_*(U+\gamma b\cdot\nabla U,U)}{|U|^2_{1,\Omega}} \tag{6.17}$$

$$= \frac{\int_\Omega \nabla U\cdot\{\varepsilon I+\gamma bb^T\}\cdot\nabla U \, d\Omega}{\int_\Omega \nabla U\cdot\nabla U \, d\Omega},$$

where I is the nxn identity matrix and b^T denotes the transpose of b. Therefore, when $n = 1$, we have $C_2 \ge \varepsilon+\gamma|b|^2$ and the bound for the ratio C_1/C_2 closely resembles (6.14) for the artificial viscosity method if, as is common, we take $\gamma = ch/|b|$, $c > 0$ constant. The methods are in fact the same when $f \equiv 0$. However, in general the two are different because the streamline upwind method modifies the source term by the addition of $(\gamma b\cdot\nabla W,f)$; this compensates for the upwinding of the term $b\cdot\nabla U$ and leads to much improved accuracy. This can be particularly noticeable in time dependent problems where the form of the mass matrix is important.

The results of this analysis are much less in accord with the observed performance of the method in higher dimensions because, from (6.17) we have $C_2 = \varepsilon$, and the ratio C_1/C_2 is consequently much larger than in one dimension. The idea behind the method is to introduce artificial viscosity only in the direction of flow (b) and this is evident from (6.17) when we interpret $\varepsilon I + \gamma bb^T$ as a tensor viscosity.

This type of analysis is, unfortunately far too crude to allow any detailed comparison of methods. Whenever boundary layers are present, we have $\nabla u = O(1/\varepsilon)$ whilst, for the numerical solution, $\nabla U = O(1/h)$ if U is to remain bounded. Thus, large changes are possible in U without affecting the error $|u-U|_{1,\Omega}$ to any significant extent (see, for example, Griffiths and Lorenz, 1978, where an attempt is made to choose the parameter γ by minimising the ratio C_1/C_2). One should, therefore, whenever possible carry out the analysis in a more suitable norm. For instance Morton (Chapter 6) makes use of norms that resemble L_2-norms when the Peclet number ($|b|/\varepsilon$) is large; Babuska and Szymczak (1982), approaching the problem from a different point of view, arrive at L_p type norms (for extensions of this work, see Szymczak and Babuska, 1983,a,b). Extensive use has also been made of the maximum norm and this work is surveyed by Ikeda (1983). All of these authors exploit the properties of nonconforming elements at some stage in their investigations illustrating the power and flexibility that these elements can offer.

REFERENCES

Babuska, I. and Aziz, A.K. (1972) Survey lectures on the mathematical foundations of the finite element method. "The Mathematical Foundations of the Finite Element Method with Applications to Partial Differential Equations", (A.K. Aziz, Ed.), Academic Press, New York, 5-359.

Bazeley, G.P., Cheung, Y.K., Irons, B. and Zienkiewicz, O.C. (1965) Triangular elements in bending, conforming and nonconforming solutions. Proc. Conf. Matrix Methods in Structural Mechanics, Wright-Patterson Air Force Base, Ohio, AFFDL-TR-66-80, 547-576.

Carey, G.F. (1976) An analysis of finite element equations and mesh subdivisions. *Comp. Meth. Appl. Mech. Engng.*, **9**, 165-179.

Christie, I., Griffiths, D.F., Mitchell, A.R. and Zienkiewicz, O.C. (1976) Finite element methods for second order differential equations with significant first derivatives. *Int. J. Num. Meth. Engng.*, **10**, 1389-1396.

Ciarlet, P.G. (1978) The Finite Element Method for Elliptic Problems. North-Holland, Amsterdam.

Crouzeix, M. and Raviart, P.A. (1973) Conforming and nonconforming finite element methods for solving the stationary Stokes equations. RAIRO R3, 33-76.

Fortin, M. and Soulie, M. (1983) A nonconforming piecewise quadratic finite element on triangles. *Int. J. Num. Meth. Engng.*, **19**, 505-520.

Griffiths, D.F. and Lorenz, J. (1978) An analysis of the Petrov-Galerkin finite element method. *Comp. Meth. Appl. Mech. Engng.*, **14**, 39-64.

Griffiths, D.F. and Mitchell, A.R. (1979) On generating upwind finite elements. "Finite Element Methods for Convection Dominated Flows", (T.J.R. Hughes, Ed.), AMD Vol. 34, ASME, New York.

Griffiths, D.F. (1979) Finite elements for incompressible flow. *Math. Meth. in the Appl. Sci.*, **1**, 16-31.

Hughes, T.J.R. and Brooks, A.N. (1979) A multi-dimensional upwind scheme with no crosswind diffusion. "Finite Element Methods for Convection Dominated Flows", (T.J.R. Hughes, Ed.), AMD Vol. 34, ASME (New York), 19-35.

Ikeda, T. (1983) Maximum Principle in Finite Element Models for Convection-Diffusion Phenomena, Mathematics Studies 76, Lecture Notes in Numerical and Applied Analysis Vol. 4, North-Holland/ Kinokuniya, Amserdam - Tokyo.

Irons, B. and Loikkanen, M. (1983) An Engineer's defence of the patch test. *Int. J. Num. Meth. Engng.*, **19**, 1391-1401.

Jones, D.S. (1980) Infinite integrals and convolution. *Proc. R. Soc., Lond.* A**371**, 479-508.

Jones, D.S. (1982) Generalised functions and their convolutions. *Proc. R. Soc. Edin.*, **91**A, 213-233.

Mitchell, A.R. and Wait, R. (1977) "The Finite Element Method in Partial Differential Equations". John Wiley and Sons, New York.

Sander, G. and Beckers, P. (1977) The influence of the choice of connectors in the finite element method. "The Mathematical Aspects of the Finite Element Method". Lecture Notes in Mathematics, Vol. 606, Springer-Verlag, Berlin, 316-340.

Shi, Zhong-ci (1983a) A convergence condition for the quadrilateral Wilson element. Manuscript, University of Frankfurt.

Shi, Zhong-ci (1983b) A note on Stummel's counterexample to the patch test. Manuscript, University of Frankfurt.

Shi, Zhong-ci (1983c) Convergence properties of two nonconforming finite elements. Manuscript, University of Frankfurt.

Shi, Zhong-ci (1983d) The generalised path test for Zienkiewicz' triangles. Manuscript, University of Frankfurt.

Shi, Zhong-ci (1983e) On the convergence properties of the quadrilateral elements of Sander and Beckers. Manuscript, University of Frankfurt.

Strang, W.G. and Fix, G.J. (1973) An Analysis of the Finite Element Method. Prentice-Hall, New York.

Stummel, F. (1979) The generalised patch test. *SIAM J. Num. Anal.*, **16**, 449-471.

Stummel, F. (1980) The limitations of the patch test. *Int. J. Num. Meth. Engng.*, **15**, 177-188.

Szymczak, W.G. and Babuska, I. (1983a) An analysis of a finite element method for convection-diffusion problems, Part 1: Quasi-optimality. Tech. Note BN-1001, University of Maryland.

Szymczak, W.G. and Babuska, I. (1983b) An analysis of a finite element method for convection-diffusion problems, Part 2: A Posteriori error estimates and adaptivity. Tech. Note BN-1002, University of Maryland.

Temam, R. (1977) Navier-Stokes Equations. North-Holland, Amsterdam.

Zienkiewicz, O.C. (1977) The Finite Element Method. 3rd ed., McGraw-Hill, London.

A-POSTERIORI ERROR ESTIMATION AND ADAPTIVE MESH REFINEMENT IN THE FINITE ELEMENT METHOD

O.C. Zienkiewicz and A.W. Craig

(Department of Civil Engineering, University College of Swansea, Swansea)

1. INTRODUCTION

The finite element method has become, since its introduction, an indispensible tool for the engineering analyst wishing to solve problems which may be posed as systems of partial differential equations. The reasons for this are well documented and the scope of the papers in these proceedings is an excellent testimony to this fact. However until recently there has been little progress in the problem of a-posteriori error estimation and the automatic optimisation of finite element meshes.

Typically, when a finite element analyst designs a mesh he is guided by his experience or that of his colleagues who may have studied similar problems in the past. When the mesh is tested on the given problem the analyst then proceeds to check for obvious errors and also checks the quality of the solution by perhaps examining stress discontinuities between elements. If the solution appears "reasonable" it is accepted, if not the analyst then tries to design a new mesh which should compensate for the deficiencies of the original mesh.

The drawbacks of this approach are obvious:

(1) There is no reliable way of judging whether the solution is acceptable or not. For this reason we want an a-posteriori error estimator, which will give an estimate of the error in the solution in some chosen norm.

(2) Assuming that we know where the error is high we need information about how best to refine the mesh, that is we need a correction indicator.

(3) We also need to eliminate the costly and wasteful intervention of the finite element analyst, who at present spends a large proportion of time designing and redesigning meshes. In other words we must make the process entirely automatic.

Thus we would like a finite element code in which the user merely has to design a mesh which is sufficient to resolve the geometry of the region in question, and then the rest is left to the computer. The computer would automatically refine the mesh in an intelligent manner (that is by introducing new degrees of freedom which produce the best possible increase in accuracy), until the error in the solution falls below some pre-specified tolerance. This code would then present the solution to the user with an estimate of the error in the solution, and would have obtained this cheaply on the best possible mesh for the problem.

As we shall see, such a code is now entirely possible using the techniques outlined below.

The structure of the remainder of this paper is as follows. After discussing the existing techniques of a-posteriori error estimation and adaptive mesh refinement we shall introduce the concepts of p-refinement and hierarchical finite elements upon which our correction indicators and error estimators are directly based. We shall explain the advantages of this formulation (many of which are valid even for non-adaptive programs) and discuss the implementation of the techniques. Finally we shall give some numerical examples to illustrate the power and the simplicity of the adaptive approach.

2. HISTORICAL SURVEY

The most obvious technique for producing an optimal mesh to solve a given set of differential equations by the finite element method is that of considering the nodal positions to be unknown (see for example McNeice and Marcal (1971)), for example if we have a self-adjoint differential equation to solve:

find $u \in V$ such that

$$a(u,v) = (f,v) \text{ for all } v \in V \tag{2.1}$$

then this may be written equivalently as

find $u \in V$ such that u minimises $I(v)$ where

$$I(v) = a(v,v) - 2(f,v) \text{ for all } v \in V \tag{2.2}$$

We may then introduce a finite element approximation to minimisation problem (2.2) with unknown nodal positions. We then solve the approximate problem by not only minimising with respect to the members of the finite element subspace approximating V but also with respect to the nodal positions defining the basis of that subspace. Such an approach however produces a non-linear matrix equation to solve which is impractical both economically and numerically. Fellipa (1976a, b) reports that the expected cost grows with the third or fourth power of the number of degrees of freedom involved.

Another technique, related to those just discussed, was introduced by Melosh and Killian (1976) and consisted of analysing the change in potential energy caused by introducing an extra degree of freedom. Peano et al. (1978a, b, 1979) proposed a similar approach based on the hierarchical approximation (see below).

Whereas the techniques outlined above gave a method of mesh refinement, none of them really gave a satisfactory error estimator. In 1975 Babuska (Babuska (1975)) produced an a-posteriori error estimator of the form (for one-dimensional elements)

$$\|e\|_{el}^2 \simeq \eta^2 = \frac{1}{12} \int_{\text{element}} r^2 \, d\Omega \tag{2.3}$$

where $\|e\|_{el}$ is the energy norm of the error over one element and r is the finite element residual

$$r = L\tilde{u} - f \tag{2.4}$$

where $\tilde{u}$ is the finite element approximation.

This was shown to converge to $\|e\|_{el}$ as the mesh parameter tended to zero. Thus by summing the contributions (2.3) we have an over-all estimate of the energy norm of the error. This technique was extended to higher dimensional problems. For more details the reader is referred to Babuska and Rheinboldt (1978a, b, 1979a, b, 1980). We can thus use this error estimator in order to refine the mesh, subdividing where the error is high, and in fact Babuska and Rheinboldt (1979c) showed that, at least in one dimension, by proceeding thus we obtain an optimal mesh, that is a mesh where the error is the same in every element. However, although this mesh becomes optimal in the limit, it is quite simple to produce examples (see, e.g., Dunavant (1980)) which show that we may not obtain the best possible decrease in error by refining the element with the highest error. This is due to the fact that the higher approximations may be orthogonal or "nearly orthogonal" to the error.

Thus we have demonstrated that there is a need not only for a reliable error estimator but also for a correction indicator which shows where best to refine the mesh. As we shall see the hierarchical finite element basis provides us with just such indicators and estimators for p-refinement, besides having many other computational advantages over the "standard" finite element basis.

3. p-REFINEMENT AND THE HIERARCHICAL FINITE ELEMENT BASIS

The concept of the hierarchical basis for a finite element approximation space was first introduced with the objective of creating conforming finite elements which would allow an easy transition from a region where the problem required higher order elements to where lower order elements were sufficient (Zienkiewicz, et al. (1970)).

The simplest way to consider a hierarchical finite element basis is to think of the basis functions as additive in the sense of terms of an infinite series, rather like a Fourier series. For example a quadratic element in one dimension (see Fig. 1) defines a standard basis of the three quadratic functions which take the unit value at one of the three nodes and are zero at the other two, however the hierarchical basis consists of the two "standard" linear basis functions defined on the element, plus a quadratic "bubble". Therefore, if we wish to increase the order of approximation of an element we simply add extra (hierarchical) basis functions to the already existing ones, we do not have to generate a completely new set of basis functions on the element. We emphasise at this point that the approximation space in both cases is exactly the same (thus we have exactly the same approximation properties) but we simply choose a different basis for that space. The concept of a hierarchical basis is easily extended to polynomials of a higher degree and also to higher dimensional problems (see Fig. 2).

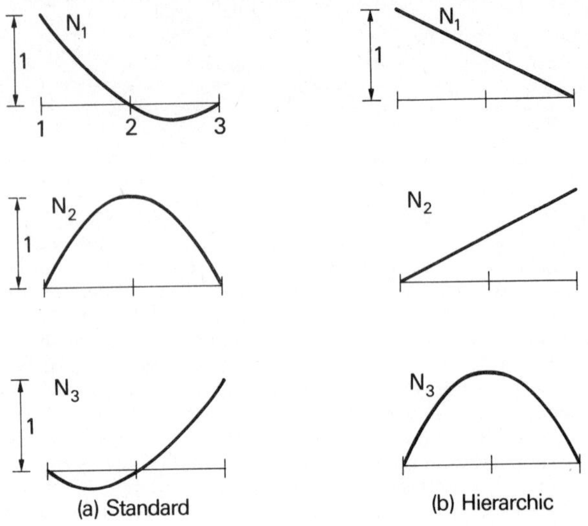

Fig. 1 Basis functions for quadratic element

To see the other advantages let us consider the following linear differential equation:

$$\begin{aligned} Lu &= f \quad \text{in } \Omega \\ Bu &= g \quad \text{on } \partial\Omega \end{aligned} \tag{3.1}$$

If we apply the standard Galerkin approximation to this problem we obtain the matrix equation:

$$K_{(n)}\underline{a}^{(n)*} = \underline{F}^{(n)} \tag{3.2}$$

where n is the number of parameters in the approximate solution. Now let us introduce m new parameters hierarchically; the discrete equations now become:

$$K_{(n+m)}\underline{a}^{(n+m)} = \underline{F}^{(n+m)} \tag{3.3}$$

or more precisely:

$$\begin{bmatrix} K_{(n)} & K_{(n,m)} \\ K_{(m,n)} & K_{(m)} \end{bmatrix} \begin{bmatrix} \underline{a}^{(n)} \\ \underline{a}^{(m)} \end{bmatrix} = \begin{bmatrix} \underline{F}^{(n)} \\ \underline{F}^{(m)} \end{bmatrix} \tag{3.4}$$

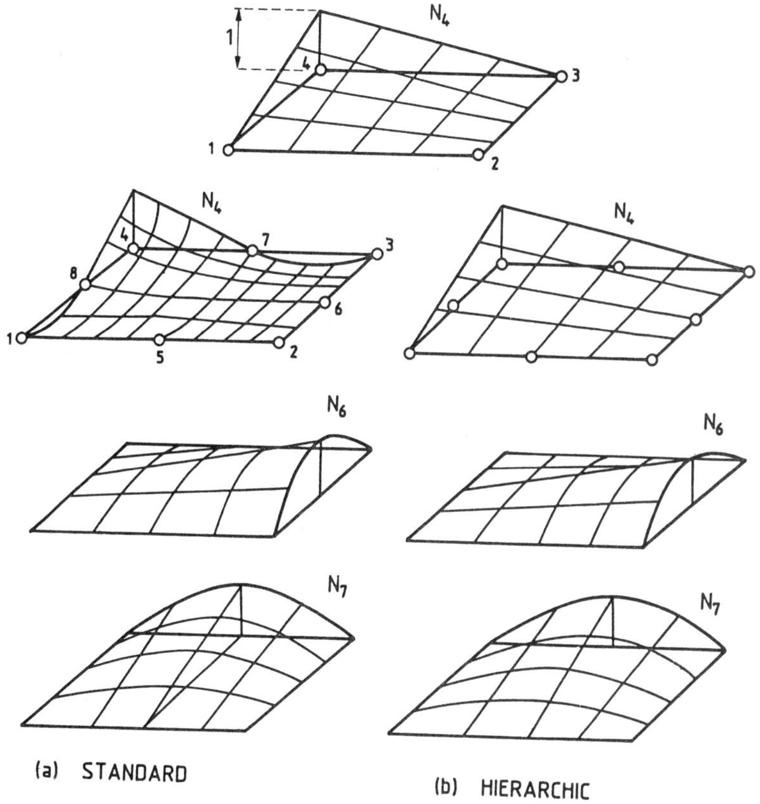

Fig. 2 Increase of polynomials in element

that is the matrix $K_{(n)}$ and the vector $\underline{F}^{(n)}$ are the same as those in equation (3.2). Thus the first advantage of using a hierarchical basis in a p-adaptive program becomes obvious.

We have preserved the matrix $K_{(n)}$ and the vector $\underline{F}^{(n)}$ from the first solution, saving a significant amount of work when calculating the new stiffness matrix and load vector. We now also have a good solution from which to start an iterative scheme to solve equation (3.4); if we solve the second system of equations in (3.4) we obtain:

$$\underline{a}^{(m)} = K_{(m)}^{-1} \left\{ \underline{F}^{(m)} - K_{(m,n)} \underline{a}^{(n)} \right\} \qquad (3.5)$$

and assuming that $\underline{a}^{(n)*} \simeq \underline{a}^{(n)}$ we obtain the approximation:

$$\underline{a}^{(n+m)} = \begin{bmatrix} \underline{a}^{(n)} \\ \\ \underline{a}^{(m)} \end{bmatrix} \simeq \begin{bmatrix} \underline{a}^{(n)*} \\ \\ K^{-1}_{(m)} (\underline{F}^{(m)} - K_{(m,n)} \underline{a}^{(n)*}) \end{bmatrix} \tag{3.6}$$

as an initial value for the new solution. If we do use an iterative method of solution then the nature of the adaptive approximation suggests that we may also gain an advantage by treating it as a multigrid. This possibility is the subject of a forthcoming paper.

We can therefore define an adaptive process in which we increase the order of elements in areas of low accuracy by adding hierarchical variables. Such a process is not only computationally convenient, but also has a faster rate of convergence than subdividing the elements (see Babuska and Dorr (1981)).

Another important advantage of hierarchical elements occurs when we consider equation conditioning. If we consider the hierarchical elements as a perturbation on the original solution, then it is obvious that the resulting stiffness matrix should be more strongly diagonally dominant. This has indeed been shown by numerical studies and it is found that with the use of hierarchical cubic elements as opposed to standard cubic elements we can expect a decrease of a factor of ten in the condition number of the matrix. Therefore, when we come to consider problems which typically produce ill conditioned matrices, for example pure traction on a slender cantilever, then there is a dramatic improvement in the solution. See, for example Zienkiewicz et al (1983b).

We also mention, in this solution, that as the hierarchical process can be considered to be a perturbation method, then it provides us with an a-posteriori estimate to the error in the approximate solution. We shall go into this in more depth later.

Thus we have shown that the use of p-hierarchical elements results in:

(1) A faster rate of convergence than the h-version

(2) The possibility of conforming mixed order interpolation

(3) The utilisation of previous solutions and computations

(4) An accurate starting value for an iterative solution procedure

(5) Improved equation conditioning

and as we shall see

(6) An a-posteriori error estimate and adaptive refinement technique.

Against this list of advantages of p-hierarchical elements the only drawback is that the coefficient of a hierarchical basis function in the finite element solution does not represent a displacement, but only a relative displacement. However as we have noted earlier there is a simple linear transformation which maps the hierarchical solution into the standard form. In fact using this transformation it would be quite simple to write a hierarchical program in which the user was never aware of the fact that hierarchical elements were being used.

We can therefore see no reason for not adopting hierarchical elements in all programs which require higher order interpolation, and in all programs which are of an adaptive nature.

4. CORRECTION INDICATORS AND ERROR ESTIMATORS

4.1 Definitions

If we are to adopt the procedures of refinement two types of question will arise. First, we shall need an indication where additional degrees of freedom will be most effective, and second, we will need a measure of error determining whether refinement is necessary.

We may define the error as

$$\underline{e} = \underline{u} - \underline{\tilde{u}} \tag{4.1}$$

and introduce the energy norm of the error:

$$\|\underline{e}\|_E^2 = \int_\Omega \underline{e}^T L\underline{e}\, d\Omega \tag{4.2}$$

This measure is of considerable importance as it can be easily related to the residual which will tell us how well the original equations are satisfied. (We neglect here any error in the approximation of the boundary conditions which can be treated similarly.) The residual obtained is

$$\underline{r} = L\underline{\tilde{u}} - \underline{f} \tag{4.3}$$

Using equation (4.1) we note that

$$L\underline{e} = L\underline{u} - L\underline{\tilde{u}} \tag{4.4}$$

and inserting equations (3.1) and (4.3) we have

$$\underline{r} = -L\underline{e} \tag{4.5}$$

Equation (4.2) can thus be rewritten as

$$\|\underline{e}\|_E^2 = -\int_\Omega \underline{e}^T \underline{r} d\Omega \tag{4.6}$$

giving one of the basic relations useful in later analysis.

4.2 Correction Indicators

As we have mentioned before if we have a reliable error estimator then it seems obvious to refine where the error is high. This however is not the case as it is possible that the next degree of refinement may be orthogonal to the error and will therefore bring no improvement in the solution, therefore we need a measure of the change in the error which would result if we introduced a particular degree of freedom. We shall call these correction indicators. Thus on the basis of these indicators calculated for all possible refinements we can decide the best way to adapt the mesh.

Let us assume that we have a solution to our problem in terms of n parameters and that we introduce a further m parameters hierarchically, then we can define the error corrected by this hierarchical refinement as

$$\underline{e}_{n+m} = \underline{\tilde{u}}^{(n+m)} - \underline{\tilde{u}}^{(n)} \tag{4.7}$$

and again using the approximation $\underline{a}^{(n)*} = \underline{a}^{(n)}$ we obtain

$$\begin{aligned} \underline{e}_{n+m} &= \underline{N}^{(m)} \underline{a}^{(n)} \\ &\simeq \underline{N}^{(m)} K_{(m)}^{-1} \left\{ \underline{F}^{(m)} - K_{(m,n)} \underline{a}^{(n)*} \right\} \end{aligned} \tag{4.8}$$

where $\underline{N}^{(m)}$ is the vector of hierarchic basis functions. In the case of a single degree of freedom introduced hierarchically we have

$$\begin{aligned} e_{n+1} &= \frac{N_{n+1} \left\{ \underline{F}^{(n+1)} - K_{(n+1,n)} \underline{a}^{(n)*} \right\}}{K_{n+1,n+1}} \\ &= \frac{-N_{n+1} \int_\Omega N_{n+1} r d\Omega}{K_{n+1,n+1}} \end{aligned} \tag{4.9}$$

So if we wish to find the energy norm of e_{n+1}

$$\|e_{n+1}\|_E^2 = - \int_\Omega e_{n+1} r d\Omega \tag{4.10}$$

by equation (4.6) and we can approximate

$$\|e_{n+1}\|_E^2 \simeq \eta_{n+1}^2$$

$$= \frac{\left\{ \int_\Omega N_{n+1} r d\Omega \right\}^2}{K_{n+1,n+1}} \tag{4.10a}$$

$$= \frac{\left\{ \underline{F}^{(n+1} - K_{(n+1,n)} \underline{a}^{(n)} \right\}^2}{K_{n+1,n+1}} \tag{4.10b}$$

While the form (4.10b) may appear more direct and has been identified as an energy error indicator by Peano et al (1979) we find that the alternative form given by (4.10a) and introduced by Zienkiewicz et al (1981), and Gago et al (1982) has a special significance since we can avoid the computation of some stiffness terms of the additional solution.

We stress once again that no additional solution is required to evaluate the correction indicators and this can be done by equation (4.10) separately for each possible new degree of freedom indicating where corrections are most desirable.

4.3 Error Estimators

Now that we have obtained correction indicators it is important that we should have a-posteriori error estimators. There are several things which we shall ask of these estimators:

(1) The error estimate should always be an overestimate of the true error

(2) If we denote the error estimator as ε and define the effectivity index as

$$\theta = \frac{\varepsilon}{\|e\|_E} \tag{4.11}$$

then we would like to have

$$\theta \to 1 \text{ as } h \to 0 \text{ or } p \to \infty \tag{4.12}$$

and θ always within reasonable bounds, say $\theta \in [1,2]$

(3) The error estimator should always be constructed on local information and should therefore also give us an estimate of local error as well as global error.

As the program user may be using the solution to check whether a design falls within certain criteria then points (1) and (2) are important. Point (1) ensures that when the error estimator has fallen within an acceptable limit then the actual error is also bounded by that limit, and point (2) guarantees that as we refine then the estimator becomes close to the error. The reason for (3) is twofold, firstly if the estimator can be constructed on an element to element basis then it will be simple to calculate and then need only be summed over the elements to obtain the total estimate. Secondly, if it can be used to estimate the error over a single element or patch of elements, then it has the important property of giving local information on the error. For example, we may be more concerned with the size of the error near a singularity, than the total error.

In principle correction indicators evaluated for all possible new degrees of freedom capable of introduction and summed should provide an accurate measure to the total energy error occurring in the solution at the refinement stage reached. Unfortunately this is impracticable and only the next degree of refinement is generally studied. If it so happens that the next basis function is orthogonal (or nearly so) to the residual and the bulk of the error lies in a higher order refinement then the error estimate will be poor - although the value of the expression (4.10) as a correction indicator continues, telling us simply that little change in the error would occur by introducing the refinement in question. Much effort has therefore gone into the problem of establishing reliable error estimators which will give accurate error values. Here work by Babuska et al (1979a) pioneered suitable estimates for linear quadrilateral elements and more recently (Zienkiewicz et al (1982), Gago et al (1981)) more elaborate expressions have been proposed for higher order elements.

We now present an error estimator first introduced by Zienkiewicz et al (1983b) which is based directly on the correction indicator (4.10). The performance of this estimator appears satisfactory and at all stages, as shown in later examples, overestimates the true error by a reasonable margin. It must be mentioned that rigorous proofs of its ultimate convergence to the true error are still lacking and that it is now proposed heuristically.

If we consider expression (4.10) we have noted that the residual may be orthogonal or nearly orthogonal to the basis function in question and that the bulk of the error may lie in higher order terms. However, by applying the Cauchy-Schwartz inequality we may say

$$\left(\int_\Omega N_{n+1} r \, d\Omega\right)^2 \leqslant \left(\int_\Omega N_{n+1}^2 \, d\Omega\right) \left(\int_e r^2 d\Omega\right) \tag{4.13}$$

and the quantity on the right hand side is always positive. Thus to estimate our error we shall evaluate

$$\varepsilon_{n+1}^2 = \frac{\left(\int_\Omega N_{n+1}^2 \, d\Omega\right)\left(\int_\Omega r^2 d\Omega\right)}{K_{n+1,n+1}} \tag{4.14}$$

for every degree of freedom available for the next refinement and obtain the total error estimator by summation.

There are several other ways in which we may extend the correction indicators to obtain an error estimator and the interested reader is referred to Zienkiewicz and Craig (1983a).

4.4 Residuals and Interelement Discontinuities

In deriving the above error estimator we have not taken account of the fact that the residual may not be well defined at element boundaries and a more complete derivation (see for example Zienkiewicz et al (1983a, b)) would lead to the expression

$$\varepsilon_{n+1}^2 = \frac{\Sigma\left(\int_{\Omega^e} N_{n+1}^2 d\Omega\right)\left(\int_{\Omega^e} r^2 d\Omega\right) + \Sigma\left(\int_{\Gamma^e} N_{n+1}^2 d\Gamma\right)\left(\int_{\Gamma^e} J^2 d\Gamma\right)}{K_{n+1,n+1}} \tag{4.15}$$

where J is the jump in gradients between elements, and Γ^e is the element boundary.

It is of interest to observe that the practical computations so far carried out show that the contribution of the interelement traction discontinuities (J) are in general a major part of the error estimator (see also Szabo (1979)). This to some extent explains the conclusion reached by many engineers that the traction jumps at interfaces present a reasonable estimate of stress errors. We mentioned that other error estimators have been derived and used with success. The one of most generality is presented in Zienkiewicz et al (1982)

$$\varepsilon^2 = \sum_{els} \left(\frac{h^2}{24kp^2} \int_{\Omega^e} r^2 d\Omega + \frac{h}{24p} \int_{\Gamma^e} J^2 d\Gamma\right) \tag{4.16}$$

in which h is the element size, p is the polynomial degree and k is a constant dependent on the problem (k=conductivity in thermal problems, $E/(1-\nu)$ in plane stress, etc.).

This explicit expression is no more easy to use and generally underestimates the error often requiring the use of corrective factors which asymptotically tend to unity as the error is reduced (Zienkiewicz et al (1981)).

5. ADAPTIVE STRATEGY AND SOLUTION PROCEDURE

Given that we have reliable correction indicators we have to decide how best to use them to obtain an accurate solution quickly and cheaply. A finite element solution will generally be started on a specified mesh. This may be deemed by the analyst to be sufficiently accurate or may only represent a very crude subdivision, machine generated with sufficient accuracy to model the boundaries only. In both cases the first step of additional computation must be the estimation of the error, and if this is found sufficiently small then the computation will be terminated. Error estimators are of primary importance here.

If the error estimate is too high, further refinement is necessary and here the correction indicators will tell us where best to refine. Various strategies are obviously possible at this point - in all the attention turning to the degrees of freedom where the correction indicators are the largest. Some of these are outlined below.

One possible approach is to always add a fixed proportion of the possible degrees of freedom in the next refinement, say 20% or 50% or 70%, corresponding to those which give the highest values of the correction indicators. This would, however, have the effect that in some cases where the error was high we may neglect to refine where it is needed, and in other cases, where the error was low we may refine where it is not needed.

Alternatively, we could decide a-priori an indicator value above which all degrees of freedom which produce a larger correction indicator would be added.

A third method would be to calculate correction indicators for several levels of refinement simultaneously and then to add the degrees of freedom corresponding to the highest indicators until we have achieved some pre-specified accuracy. For example, if our error estimator told us that we had an error of 20% and we wished only to have an error of 5%, then we would add enough degrees of freedom in order to give us a decrease of 15% (estimated by summing the indicators). This seems to be the best approach and is the subject of current research.

For the purposes of this paper however we adopt a simpler approach which falls between the first two outlined above:

we decide on a parameter γ such that

$$\gamma \in (0,1) \tag{5.1}$$

and add all the degrees of freedom corresponding to correction indicators η such that

$$\eta > \gamma \eta_{max} \tag{5.2}$$

thus the parameter γ controls the speed of convergence, if γ is zero then we add all possible degrees of freedom, if γ is one then we add none. The choice of $\gamma = 0$ has been adopted in one of the recently launched commercial programs.

In general most of the correction indicators are of a value less than a tenth of the maximum indicator and in fact a choice of $\gamma = 0.1$ appears to give the largest decrease in error for a given amount of computer time (Zienkiewicz and Craig (1983a)). So the adaptive technique is cheaper than using complete refinement.

Thus the adaptive process is justified in terms of cost alone. There are, however, other very strong reasons to use the adaptive approach, especially when used as a p-adaptive hierarchical process. Firstly, the same accuracy is achieved using fewer degrees of freedom than in the complete refinement process, therefore the storage requirements in a computer are much less. This is particularly important with the advent of microcomputers, and brings the power of the finite element method within the scope of the engineer with limited computing facilities. Secondly, we mention again the improvement in equation conditioning which, as well as being important when solving problems which are typically ill-conditioned, is also important when using a small machine of limited accuracy.

More importantly, we expect the adaptive process to lead to a mesh on which the errors are equally distributed. We may, for example achieve 5% accuracy using a complete refinement process, however as this accuracy is in terms of the energy norm and thus averaged over the mesh we may in fact have 1% accuracy where the stress concentrations are low and 10% accuracy where they are high, and it is precisely where the stress concentrations are high that we need the most accurate solution.

In the adaptive process we expect to obtain an optimal mesh in the limit. If the energy norm error is 5% then we expect it to be near 5% in every element in the mesh. Indeed, due to the local nature of the error estimators we may check this. Babuska and Rheinboldt (1979c) have in fact shown that we obtain an optimal mesh for an h-adaptive process. These facts alone would justify an adaptive process, even if it were not cost-effective.

6. NUMERICAL EXAMPLES

In this section we shall demonstrate the efficiency of the proposed error estimators for two different examples: firstly an L-shaped region with a corner singularity (see Fig. 3) and secondly the same problem with a temperature field imposed. For further examples the reader is referred to Zienkiewicz et al (1983a,b).

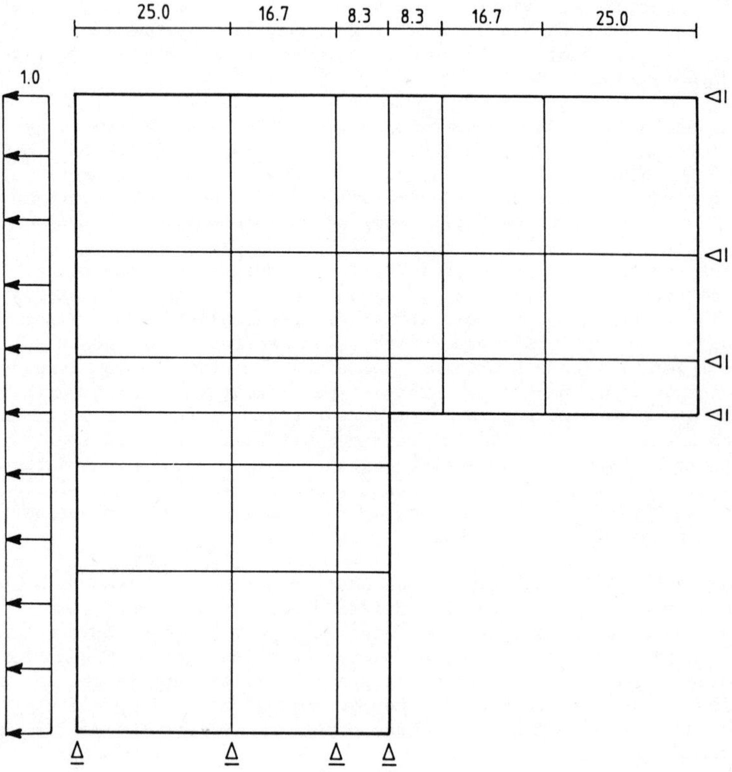

Fig. 3a L-shaped region, initial mesh and load conditions

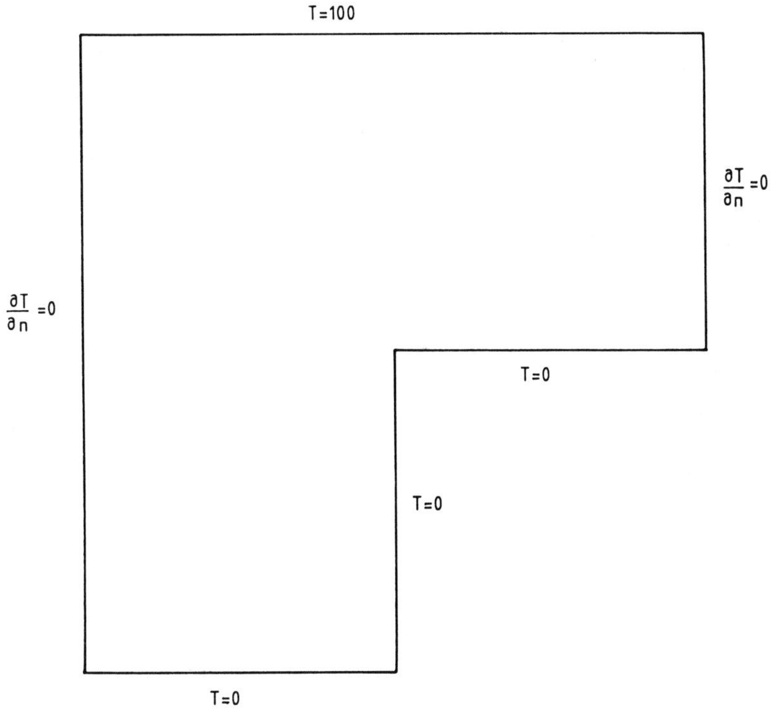

Fig. 3b L-shaped region, temperature boundary conditions

In both cases we present the results arising from the use of $\gamma = 0.1$, which as we mentioned in the last section was the most efficient choice of refinement parameter. In tables 1 and 2 we show the results for the two examples. As we can see, the new error estimator consistently overestimates the error to an acceptable degree. Also, in both of these examples the new error estimator converges to a value of around 1.4. We may therefore conjecture that an estimator of the form

$$\varepsilon_{n+1}^{*2} = \frac{1}{(1.3)^2}\,\varepsilon_{n+1}^{2} \tag{6.1}$$

will give a close overestimate of the error in the solution, and we also present results for this form in tables 1 and 2. It is also clear, at least for the examples which we have given, that our adjusted estimator (6.1) has all of the properties which we required of error estimators.

Table 1

Efficiency of Error Indicators for L-shaped Domain

est1 = Zienkiewicz et al 1982.

est2 = Zienkiewicz et al 1983a, b.

est3 = est2/1.3

$\gamma = 0.1$

(figure in parenthesis is the estimated error expressed as a percentage of $\|u\|$)

solution number	Ndof	exact $\|e\|$	est1	est2	est3	θ_1	θ_2	θ_3
1	72	1.44 (25.8)	1.00 (17.9)	2.45 (43.9)	1.88 (33.7)	0.69	1.70	1.31
2	120	0.56 (10.0)	0.39 (7.0)	0.83 (14.9)	0.64 (11.5)	0.70	1.48	1.14
3	171	0.22 (3.9)	0.14 (2.5)	0.31 (5.6)	0.24 (4.3)	0.64	1.42	1.09
4	213	0.11 (2.0)	0.09 (1.6)	0.15 (2.7)	0.12 (2.2)	0.75	1.40	1.07

Table 2

Efficiency of Error Indicators for L-shaped domain with temperature field

est1 = Zienkiewicz et al 1982.

est2 = Zienkiewicz et al 1983a, b.

est3 = est2/1.3

$\gamma = 0.1$

(figure in parenthesis is the estimated error expressed as a percentage of $\|u\|$)

solution number	Ndof	exact $\|e\|$	est1	est2	est3	θ_1	θ_2	θ_3
1	72	2.01 (28.1)	1.35 (18.9)	3.45 (48.2)	2.65 (37.1)	0.67	1.72	1.32
2	122	0.78 (10.9)	0.53 (7.4)	1.16 (16.2)	0.90 (12.6)	0.68	1.49	1.15
3	175	0.33 (4.6)	0.23 (3.2)	0.47 (6.6)	0.36 (5.0)	0.70	1.43	1.10
4	211	0.17 (2.4)	0.13 (1.8)	0.24 (3.4)	0.18 (2.5)	0.75	1.41	1.08

REFERENCES

Babuska, I (1975) The self-adaptive approach in the finite element method. In MAFELAP 2 (J.R. Whiteman, Ed.), Academic Press, London.

Babuska, I. and Dorr, M.R. (1981) Error estimates for the combined h and p versions of the finite element method. *Num. Math.*, **25**, 257-277.

Babuska, I. and Rheinboldt, W.C. (1978a) Error estimates for adaptive finite element computations. *SIAM J. Num. Anal.*, **15**, No. 4.

Babuska, I. and Rheinboldt, W.C. (1978b) A-posteriori error estimates for the finite element method. *Int. J. Num. Meth. Engng.*, **12**, 1597-1615.

Babuska, I. and Rheinboldt, W.C. (1979a) Adaptive approaches and reliability estimates in finite element analysis. *Comp. Meth. App. Mech. Engng.*, **17/18**, 519-540.

Babuska, I. and Rheinboldt, W.C. (1979b) A-posteriori error analysis of finite element solutions for one-dimensional problems. Inst. for Comp. Math. and App., technical report ICMA-79-11, University of Pittsburgh.

Babuska, I. and Rheinboldt, W.C. (1979c) Analysis of optimal finite element meshes in R. *Math. Comput.*, **30**.

Babuska, I. and Rheinboldt, W.C. (1980) Reliable error estimation and mesh adaptation for the finite element method. In Computational Methods in Nonlinear Mechanics, (J.T. Oden, Ed.), 67-108.

Dunavant, D.A. (1980) Local a-posteriori indicators of error for the p-version of the finite element method. Report no. WU/CCM-80/1, Centre for Computational Mathematics, Washington University.

Fellipa, C.A. (1976a) Numerical experiments in finite element grid optimisation by direct energy search. *App. Math. Mod.*, **1**.

Fellipa, C.A. (1976b) Optimisation of finite element grids by direct energy search. *App. Math. Mod.*, **1**.

Gago, J.P. de S.R., Kelly, D.W. and Zienkiewicz, O.C. (1981) Adaptive finite element schemes and a-posteriori error analysis, an explanation of the alternatives. Proc. 3rd World Congress on Finite Element Methods, Los Angeles (Robinson, Ed.).

McNeice, B.M. and Marcal, F.V. (1971) Optimisation of finite element grids based on maximum potential energy. Technical Report no. 7, Brown University Providence.

Mellosh, R.J. and Killian, B.E. (1976) Finite element analysis to obtain a pre-specified accuracy. Proc. 3rd Nat. Cong. on Computing in Structures.

Peano, A., Riccioni, R., Passini, A. and Sardella, L. (1978a) Adaptive approximations in finite element structural analysis. ISMES, Bergamo, Italy.

Peano, A. and Riccioni, R. (1978b) Automated discretisation error control in finite element analysis. 2nd World Congress in Finite Element Methods.

Peano, A., Fanelly, M., Riccioni, R. and Sardella, L. (1979) Self-adaptive convergence at the crack tip of a dam buttress. Int. Conf. on Numerical Method. in Fracture Mechanics, Swansea.

Szabo, B.A. (1979) Some recent developments in finite element analysis, *Comput. Maths. Appl.*, **5**, 99-115.

Zienkiewicz, O.C. and Craig, A.W. (1983a) Adaptive mesh refinement and a-posteriori error estimation for the p-version of the finite element method. Given at Workshop on Adaptive Methods for Partial Differential Equations, University of Maryland, February 1983. Proc. to be published in SIAM (I. Babuska and J. Chandra, Eds.).

Zienkiewicz, O.C., Gago, J.P. de S.R. and Kelly, D.W. (1983b) The hierarchical concept in finite element analysis. *Comp. and Structures*, **16**, 53-65.

Zienkiewicz, O.C., Kelly, D.W., Gago, J.P. de S.R. and Babuska, I. (1982) Hierarchical finite element approaches, adaptive refinement and error estimates. Proc. MAFELAP 1981, (J.R. Whiteman, Ed.), Academic Press, London.

Zienkiewicz, O.C., Irons, B.M., Scott, F.E. and Campbell, J.S. (1970) High speed computing of elastic structures. Proc. of the Symposium of The International Union of Theoretical and Applied Mechanics, Liege.

FINITE ELEMENT METHODS FOR NON-SELF-ADJOINT ELLIPTIC AND FOR HYPERBOLIC PROBLEMS: OPTIMAL APPROXIMATIONS AND RECOVERY TECHNIQUES

K.W. Morton

(Department of Mathematics, University of Reading)
(Now at Oxford University Computing Laboratory)

1. INTRODUCTION

In this article we shall consider the progress that has been made in extending and developing the finite element method so that it may be applied to much wider classes of problems than that for which it was originally developed. Within the context of stress problems, where engineers originated many of the early ideas, the method could be based on an extremal principle - for the strain energy. Mathematically, such principles always lead to self-adjoint elliptic problems: alternative principles are therefore needed for more general problems. Some form of variational principle or weak formulation is usually available but the way forward is less clear-cut and the results can often be disappointing.

Just as with finite difference methods, it is not too difficult to devise approximation schemes for quite general problems which will converge as some mesh parameter tends to zero: one may even achieve the optimal order of convergence. But the marvellous economy and robustness on coarse meshes, which were key features of finite element methods in their original context, will be lost unless rather special steps are taken to preserve the optimal approximation properties and super-convergence properties which lie behind them.

How this may be done is the prime theme linking the three lectures on which this article is based. We shall concentrate on essential ideas, giving references where more details and more specific applications may be found. Much of the author's thinking is based on work with students and colleagues which is gratefully acknowledged and which can be only partially referenced.

To fix our notation, let us begin by recalling the main properties of finite element approximations to the following self-adjoint problem for the real scalar function u: in classical form,

$$Lu = f \text{ in } \Omega \tag{1.1a}$$

$$u = g \text{ on } \Gamma_D, \ \partial u/\partial n = 0 \text{ on } \Gamma_N, \tag{1.1b}$$

where we will assume Ω is a bounded, open region of $\mathbb{R}^2$, with boundary composed of the two non-intersecting portions Γ_D and Γ_N, and L is a second order linear elliptic operator. In weak form, we write this:- find $u \in H^1_E$ such that

$$A(u,w) = (f,w) \qquad \forall\, w \in H^1_{E_0}, \tag{1.2}$$

where $(\cdot,\cdot)$ denotes the usual L_2 inner product, H^1_E denotes elements of the usual Sobolev space $H^1(\Omega)$ which satisfy the essential boundary condition $u = g$ on Γ_D and $H^1_{E_0}$ denotes the associated subspace of elements satisfying the corresponding homogeneous condition; here the symmetric form $A(u,w)$ can be obtained formally from (Lu,w) by integrating by parts and applying the boundary conditions $\partial u/\partial n = 0$ on Γ_N and $w = 0$ on Γ_D. We define the

$$\text{energy norm, } \|v\|_A := \{A(v,v)\}^{\frac{1}{2}}, \tag{1.3}$$

its positive definiteness following from the coercivity of $A(\cdot,\cdot)$, i.e. the ellipticity of L (see below). In many fields of application, the basic physical principles are best expressed in the extremal principle form:-

find $u \in H^1_E$ to

$$\text{minimise} \quad A(v,v) - 2(f,v) \quad \text{over } H^1_E. \tag{1.4}$$

Variation of this expression clearly leads directly to (1.2).

Suppose now that Ω can be precisely divided into elements, on which a finite element basis $\{\phi_i(\underline{x})\}$ can be constructed, and g is such that a conforming trial space S^h_E can be defined as:-

$$H^1_E \supset S^h_E := \{U(\underline{x}) = \Sigma\, U_j\phi_j(\underline{x}) \mid U = g \quad \text{on} \quad \Gamma_D\}, \tag{1.5a}$$

together with a corresponding test space S^h_0 given by

$$H^1_{E_0} \supset S^h_0 := \{V(\underline{x}) = \Sigma\, V_j\phi_j(\underline{x}) \mid V = 0 \quad \text{on} \quad \Gamma_D\}. \tag{1.5b}$$

Then carrying out the minimisation (1.4) over S^h_E leads to the Galerkin approximation U to u given by:-

$U \in S^h_E$ such that

$$A(U,W) = (f,W) \qquad \forall\, W \in S^h_0. \tag{1.6}$$

Because of the conforming property, (1.2) is true with W substituted for w so that subtracting (1.6) from the result gives

$$A(u-U,W) = 0 \qquad \forall\, W \in S_0^h. \qquad (1.7)$$

That is, the error u-U is orthogonal to the test space S_0^h in the inner product defined by $A(\cdot,\cdot)$. Because of the symmetry of $A(\cdot,\cdot)$, the optimal approximation property of U follows immediately:-

$$\|u-U\|_A = \inf_{V \in S_E^h} \|u-V\|_A \,. \qquad (1.8)$$

This is the key property of finite element methods which we wish to carry over into more general problems: first to non-self-adjoint elliptic operators L, which we shall do in section 2 mainly by reference to diffusion-convection problems; and then to hyperbolic problems which we shall treat in section 4.

To appreciate the significance of (1.8), let us suppose L is the Laplacian operator so that $A(u,w) = (\underline{\nabla} u, \underline{\nabla} w)$ and (1.8) becomes

$$\int_\Omega |\,\underline{\nabla}(u-U)\,|^2 d\Omega = \inf_{V \in S_E^h} \int |\,\underline{\nabla}(u-V)\,|^2 d\Omega. \qquad (1.9)$$

Suppose also that the simplest conforming elements are used, that is piecewise linear elements over triangles, with the U_j in (1.5a) corresponding to the nodal values at the triangle vertices. Then $\underline{\nabla} U$ is piecewise constant and (1.9) shows that it is the best such piecewise constant approximation to $\underline{\nabla} u$ in the least squares sense. For practical calculations it is very often the field $\underline{\nabla} u$ rather than the potential u which is of most interest and the piecewise constant approximation can only be at best first order accurate at most points. However, we shall see in section 3 that a second order approximation can be constructed from U because of the superconvergence properties implied by (1.9): unlike the case of bilinear elements on rectangles, there are no points (such as the centroid) of each triangle where $\underline{\nabla} U$ is second order accurate and a simple recovery procedure is needed; moreover, this construction also depends on the arrangement of the triangles such that exactly six have each vertex in common.

2. NON-SELF-ADJOINT ELLIPTIC PROBLEMS

In weak form and for second order operators, these can be written as in (1.2): find $u \in H_E^1$ such that

$$B(u,w) = (f,w) \qquad \forall\, w \in H_{E_0}^1 , \qquad (2.1)$$

where now $B(\cdot,\cdot)$ is an unsymmetric bilinear form. The theory is simplest if we have homogeneous boundary conditions, while of course we wish to allow for inhomogeneous Dirichlet data in practice, as in (1.1b). Thus suppose the boundary and the Dirichlet data are smooth enough that g is the restriction to Γ_D of a function $G \in H^1(\Omega)$: it is sufficient, for example, that the boundary is locally Lipschitz continuous and $g \in L_2(\Gamma_D)$ - see Ciarlet (1978) or any similar text. Then (2.1) can be rewritten as: find $u^0 \in H^1_{E_0}$ such that

$$B(u^0,w) = (f^0,w) := (f,w) - B(G,w) \qquad \forall\, w \in H^1_{E_0}, \tag{2.2}$$

where $u = u^0 + G$ and we have introduced f^0 which lies in the dual space of $H^1_{E_0}$, to be denoted by $H^{-1}_{E_0}$. Existence and uniqueness of the solution to (2.2) results from the following lemma.

<u>Lax-Milgram Lemma</u>. Suppose $B(\cdot,\cdot)$ is a bilinear form on $H^1(\Omega) \times H^1(\Omega)$, where $H^1(\Omega)$ is equipped with the norm $\|\cdot\|_B$, and it is

(i) <u>continuous</u>, i.e. $\exists$ a constant K such that

$$|B(v,w)| \le K\|v\|_B\|w\|_B \qquad \forall\, u,w \in H^1_{E_0}, \tag{2.3a}$$

and (ii) <u>coercive</u>, i.e. $\exists$ a constant $\alpha > 0$ such that

$$B(u,u) \ge \alpha\|u\|_B^2 \qquad \forall\, u \in H^1_{E_0}. \tag{2.3b}$$

Then there exists a unique solution $u^0 \in H^1_{E_0}$ to (2.2) for every $f^0 \in H^{-1}_{E_0}$. (Note that if $\|\cdot\|_B$ is taken from the symmetric part of $B(\cdot,\cdot)$, i.e. $2(u,w)_B = B(u,w) + B(w,u)$ and $\|u\|_B^2 = (u,u)_B$, then $\alpha = 1$ and K measures the asymmetry of $B(\cdot,\cdot)$.)

Although we no longer have an extremal principle from which the Galerkin approximation can be derived, such an approximation can be defined directly from (2.1) or (2.2). The latter actually allows a larger class of g to be treated than that based on (1.5a). Thus we replace this definition by

$$H^1_E \supset S^h_E := \{U = G + V \,|\, V \in S^h_0\} \tag{2.4}$$

and then have the discrete problem:- find $U \in S_E^h$ such that

$$B(U,W) = (f,W) \qquad \forall\, W \in S_O^h \tag{2.5}$$

and, just as with (1,7), we have

$$B(u-U,W) = 0 \qquad \forall\, W \in S_O^h. \tag{2.6}$$

Unfortunately, instead of (1.8) all we can now prove from (2.3) is the following: for any $V \in S_E^h$, setting $W = U-V$ in (2.6),

$$\alpha\|u-U\|_B^2 \le B(u-U,u-U) = B(u-U,u-V)$$

$$\le K\|u-U\|_B \|u-V\|_B.$$

Hence we have

$$\|u-U\|_B \le (K/\alpha) \inf_{V \in S_E^h} \|u-V\|_B. \tag{2.7}$$

This means that, while the same order of convergence in the norm $\|\cdot\|_B$ is obtained as in the self-adjoint case, the constant in the asymptotic rate may be greatly increased: moreover, the crucial superconvergence properties are lost.

Diffusion-convection problems

These form an important class of practical problems which illustrate the difficulties. They are of the following form:

$$-\underline{\nabla}\cdot(a\underline{\nabla}u - \underline{b}u) = f \text{ in } \Omega \tag{2.8a}$$

$$u = g \text{ on } \Gamma_D, \qquad \partial u/\partial n = 0 \text{ on } \Gamma_N. \tag{2.8b}$$

Here a is a (positive) diffusion coefficient and $\underline{b}$ a convective velocity field which we shall assume is incompressible (i.e. $\underline{\nabla}\cdot\underline{b} = 0$). The corresponding bilinear form is

$$B(u,w) := (a\underline{\nabla}u, \underline{\nabla}w) + (\underline{\nabla}\cdot(\underline{b}u),w) \tag{2.9}$$

and to ensure its coercivity we assume that Γ_D includes all points of the boundary on which $\underline{b}\cdot\underline{n} < 0$, so that u is prescribed on the inflow boundary. Indeed, it is easy to see that

$$B(u,u) = (a\underline{\nabla}u,\underline{\nabla}u) + \tfrac{1}{2}\int_{\Gamma_N} (\underline{b}\cdot\underline{n})u^2 ds \qquad \forall\, u \in H^1_{E_0}. \tag{2.10}$$

Thus, if $a\in C^0(\Omega)$ and $\underline{b}\in[H^1(\Omega)]^2$, $B(\cdot,\cdot)$ clearly satisfies all the hypotheses of the Lax-Milgram Lemma.

Suppose we introduce the symmetric form

$$B_1(u,w) := (a\underline{\nabla}u,\underline{\nabla}w), \tag{2.11}$$

with associated norm $\|\cdot\|_{B_1}$, and let U_1^* be the best fit to u in this norm from the trial space S_E^h: that is,

$$B_1(u-U_1^*,W) = 0 \qquad \forall\, W \in S_0^h. \tag{2.12}$$

Then for the Galerkin approximation U given by (2.5) we have, comparing (2.10) with (2.11),

$$\begin{aligned}\|u-U\|_{B_1}^2 &\le B(u-U,u-U) = B(u-U,u-U_1^*)\\ &= B_1(u-U,u-U_1^*) + (\underline{b}\cdot\underline{\nabla}(u-U),u-U_1^*)\\ &\le \|u-U\|_{B_1}\{\|u-U_1^*\|_{B_1} + \max_{\Omega}(|\underline{b}|/a)\|a^{\frac{1}{2}}(u-U_1^*)\|\}.\end{aligned} \tag{2.13}$$

From the Aubin-Nitsche duality argument (see earlier chapter on self-adjoint problems) it is easy to deduce that there exists a constant K, independent of $\underline{b}$, such that

$$\|a^{\frac{1}{2}}(u-U_1^*)\| \le Kh\|u-U_1^*\|_{B_1}. \tag{2.14}$$

where h is the largest diameter of any element. It therefore follows that

$$\|u-U\|_{B_1} \le [1 + Kh\max_{\Omega}(|b|/a)]\|u-U_1^*\|_{B_1}. \tag{2.15}$$

This is sharper than (2.7), showing the dependence on element size through the important dimensionless parameter, the mesh Péclet number bh/a.

A useful simple test problem in one dimension is:

$$-au'' + bu' = f \text{ on } (0,1) \tag{2.16a}$$

$$u(0) = 0, \quad u(1) = 1, \tag{2.16b}$$

where a and b are positive constants. For $f = 0$, the solution is easily seen to be

$$u(x) = (e^{bx/a} - 1)/(e^{b/a} - 1): \tag{2.17}$$

piecewise linear elements on a uniform mesh of size h give the Galerkin equations for $j = 1,2,\ldots,J-1$ with $Jh = 1$

$$-\delta^2 U_j + (bh/a)\Delta_0 U_j = 0, \tag{2.18}$$

where we have used the usual finite difference notation $\delta^2 U_j := U_{j+1} - 2U_j + U_{j-1}$, $\Delta_0 U_j := \frac{1}{2}(U_{j+1} - U_{j-1})$; these have the solution

$$U_j = (\mu_0^j - 1)/(\mu_0^J - 1), \qquad \mu_0 = (2+bh/a)/(2-bh/a). \tag{2.19}$$

Clearly, the approximation for $bh/a > 2$ exhibits oscillations which bear no relation to the exponential solution (2.17) and the error bound (2.15) is seen to be quite realistic, the K in this case being calculated as $1/\pi$.

These spurious oscillations are a well-known result of the central differences yielded by the Galerkin method in (2.18). In difference methods they are overcome by using upwind differencing, replacing $\Delta_0 U_j$ by $\Delta_- U_j := U_j - U_{j-1}$ or by a weighted average of the two. The best-known scheme is that due to Allen and Southwell (1955) which with the average $(1-\xi)\Delta_0 + \xi\Delta_-$ can be written as

$$-[1+\tfrac{1}{2}\xi(bh/a)]\delta^2 U_j + (bh/a)\Delta_0 U_j = 0; \tag{2.20}$$

for the choice

$$\xi = \coth(\tfrac{1}{2}bh/a) - (\tfrac{1}{2}bh/a)^{-1} \tag{2.21}$$

we obtain the so-called exponentially-fitted scheme which gives exact nodal values for this model problem.

Petrov-Galerkin methods

The first finite element methods to overcome the deficiencies of the Galerkin method followed similar lines and used different weight functions from the trial functions ϕ_j with a view to generating these upwind difference schemes. In general, for a Petrov-Galerkin method we introduce a test space T_0^h different from but with the same dimension as the S_0^h of (1.5b): with basis functions $\psi_j(\underline{x})$, usually over the same elements, we have

$$H^1_{E_0} \supset T_0^h := \{V(\underline{x}) = \Sigma\, V_j\psi_j(\underline{x}) \mid V = 0 \text{ on } \Gamma_D\}. \tag{2.22}$$

Then the Galerkin method of (2.5) is generalised to find $U \in S_E^h$ such that

$$B(U,W) = (f,W) \qquad \forall\, W \in T_0^h. \tag{2.23}$$

The important question is "how should T_0^h be chosen for a given trial space S_E^h?". In particular, can it be done satisfactorily without reference to the upwind difference schemes that it might be induced to give on a regular mesh? There is a large literature concerned with the development of Petrov-Galerkin methods for diffusion-convection problems and most approaches make some use of this idea: see, for example, the conference proceedings Hughes (1979). We shall however follow Barrett and Morton (1980, 1981) and Morton (1982) in basing our approach on symmetrizing the bilinear form $B(\cdot,\cdot)$.

Suppose $B_S(\cdot,\cdot)$ is a symmetric bilinear form giving an inner product and norm $\|\cdot\|_{B_S}$ on $H^1_{E_0}$ with respect to which $B(\cdot,\cdot)$ satisfies the hypotheses (2.3a) and (2.3b) of the Lax-Milgram Lemma. Then, for any fixed w, $B(u,w)$ is a bounded linear functional of u and by the Riesz Representation Theorem can be written $B_S(u,Rw)$ where Rw is an element of $H^1_{E_0}$: indeed, by the linearity of $B(\cdot,\cdot)$ and (2.3), R is a linear operator on $H^1_{E_0}$ and we can write

$$B(u,w) = B_S(u,Rw) \qquad \forall u,\ w \in H^1_{E_0}. \tag{2.24}$$

Effectively, R is a symmetrizer for $B(\cdot,\cdot)$. Note too that the coercivity condition (2.3b) ensures that R is invertible on $H^1_{E_0}$. Now for the Petrov-Galerkin approximation given by (2.23) we have, corresponding to (2.6),

$$B(u-U,W) = 0 \qquad \forall W \in T_0^h \tag{2.25}$$

which by (2.24) we can now write as

$$B_S(u-U,RW) = 0 \qquad \forall W \in T_0^h. \tag{2.26}$$

The extent to which this leads to the optimal approximation property of (1.7) and (1.8) is then given by the following theorem.

Theorem (Morton, 1982). Suppose the test space T_0^h has the same dimension as S_0^h and that there exists a constant $\Delta\in[0,1)$ such that

$$\inf_{W\in T_0^h} \|V-RW\|_{B_S} \le \Delta\|V\|_{B_S} \qquad \forall V \in S_0^h. \tag{2.27}$$

Then there exists a unique solution $U^0 \in S_0^h$ to (2.23) for every $f^0 \in H^{-1}_{E_0}$ and the error between U^0 and the solution u^0 of (2.2) satisfies

$$\|u^0 - U^0\|_{B_S} \le (1-\Delta^2)^{-\frac{1}{2}} \inf_{V \in S_0^h} \|u^0-V\|_{B_S}. \tag{2.28}$$

The effect of inhomogeneous Dirichlet data is dealt with as described at the beginning of this section. The result (2.28) is also shown in the above reference to include and be somewhat sharper than that of the Generalised Lax-Milgram theorem of Babuska and Aziz (1972).

This theorem in principle enables an error bound to be calculated for any Petrov-Galerkin method, so long as sufficient knowledge of R is available for the approximation result (2.27) to be obtained. Note that this result holds for all data f and if (2.27) is sharp then so is (2.28): however for specific data (2.28) may not be particularly sharp.

This framework also allows two alternative approaches to the task of constructing effective Petrov-Galerkin methods. The first, conventional, approach is to construct basis functions ψ_i of T_0^h in such a way that the constant Δ in (2.27) is small but also so as to have local support so that the stiffness matrix $B(\phi_j,\psi_i)$ resulting from their substitution in (2.23) is easily evaluated and has small bandwidth. Note however that this matrix will be unsymmetric and the solution of (2.23) correspondingly more difficult than that of the Galerkin equations for a self-adjoint problem. The alternative approach is based on the ideal test functions ψ_i^* which are the solution of the equations

$$R\psi_i^* = \phi_i \qquad \forall \phi_i \in S_0^h. \tag{2.29}$$

Use of this test space would give $\Delta = 0$ in (2.27) and yield the optimal approximation to u^0 in the $\|\cdot\|_{B_S}$ norm, i.e. that which achieves the infimum on the right-hand side of (2.28): so we would have completely achieved our original objective. Moreover, the Petrov-Galerkin equations (2.23) could then be written for this $U^* \in S_0^h$ as

$$B_S(U^*,\phi_i) = (f,\psi_i^*) - B(G,\psi_i^*) \qquad \forall \phi_i \in S_0^h, \tag{2.30}$$

where as before G is the extension of the boundary data. These equations have the practical advantage of being symmetric: indeed they are the same as the Galerkin equations for a self-adjoint problem corresponding to $B_S(\cdot,\cdot)$. What remains is to approximate ψ_i^* sufficiently well for the right-hand side of (2.30) to be calculated to adequate accuracy. This is a linear functional of ϕ_i which we will write as

$$F^*(\phi_i) := (f,R^{-1}\phi_i) - B(G,R^{-1}\phi_1) \qquad \forall \phi_i \in S_0^h. \tag{2.31}$$

Suppose now that this is approximated by $F(\phi_i)$, an approximation for which we can establish the error bound

$$|F(V) - F^*(V)| \leqq \delta_F \|V\|_{B_S} \qquad \forall V \in S_0^h. \tag{2.32}$$

Then the corresponding approximation U^0 to u^0 is given by

$$B_S(U^0,\phi_i) = F(\phi_i) \qquad \forall V \in S_0^h \tag{2.33}$$

and clearly satisfies

$$\|U^*-U^0\|^2_{B_S} = |F^*(U^*-U^0) - F(U^*-U^0)| \leqq \delta_F \|U^*-U^0\|_{B_S}. \tag{2.34}$$

There results, using the optimality property of U*, the error bound

$$\|u^0-U^0\|^2_{B_S} \leqq \|u^0-U^*\|^2_{B_S} + \delta_F^2. \tag{2.35}$$

That is, we have a term added to the optimal error estimate as compared with the multiplicative factor of (2.28); and, moreover, this term can be determined for specific data. Note that a data independent error bound may also be obtained if required but that it will still be additive: for example, suppose $F(\phi_i)$ is computed through a linear operator $T: S_0^h \to H^1_{E_0}$, approximating R^{-1},

$$\begin{aligned} F(\phi_i) &:= (f,T\phi_i) - B(G,T\phi_i) \\ &= B(u^0,T\phi_i) = B_S(u^0,RT\phi_i); \end{aligned} \tag{2.36}$$

then we have

$$\begin{aligned} |F(V) - F(V^*)| &= |B_S(u^0, RTV - V)| \\ &\leqq \|[I-(RT)^*]u^0\|_{B_S} \|V\|_{B_S} \qquad \forall V \in S_0^h \end{aligned} \tag{2.37}$$

where $(RT)^*$ is the adjoint of RT in the inner product $B_S(\cdot,\cdot)$.

Application to diffusion-convection

We conclude this section by outlining the application of these ideas to the diffusion-convection problems given by (2.8) and (2.16): more details can be found in Barrett and Morton (1983) and the references therein. The operator in (2.8a) can be factored to give

$$L_1^* L_2 u = f, \tag{2.38}$$

where

$$L_1 v := a^{\frac{1}{2}}\underline{\nabla} v, \; L_2 v := a^{\frac{1}{2}}\underline{\nabla} v - (\underline{b}/a^{\frac{1}{2}})v$$

and L_1^* is the formal adjoint of L_1. This suggests two distinct symmetric bilinear forms, one based on L_1 and one on L_2. We have already introduced the former and called it $B_1(\cdot,\cdot)$ in (2.11): we now define the second choice by

$$B_2(v,w) := (a\underline{\nabla}v,\underline{\nabla}w) + ((b^2/a)v,w) \tag{2.39a}$$

$$= (L_2v,L_2w) + \int_\Gamma (\underline{b}\cdot\underline{n})vwds. \tag{2.39b}$$

Let us denote by R_1 and R_2 the Riesz representer in (2.24) and by Δ_1 and Δ_2 the smallest constant in the approximation estimate (2.27) in these two cases. Then a little manipulation shows that these constants are given by the following discrete minimisation problems:

$$1 - \Delta_m^2 = \min_{\underline{V}} \left\{\frac{\underline{V}^T B^T A^{-1} B\underline{V}}{\underline{V}^T C\underline{V}}\right\}, \quad m = 1,2 \tag{2.40}$$

where the matrices A,B,C have components given by

$$A_{ij} = B_m(R_m\psi_j,R_m\psi_i)$$

$$B_{ij} = B_m(\phi_j,R_m\psi_i) = B(\phi_j,\psi_i)$$

$$C_{ij} = B_m(\phi_i,\phi_j)$$

and $\underline{V}^T = \{V_1,V_2,\ldots,V_N\}, \quad V = \sum_1^N V_j\phi_j \in S_0^h.$

For the model problem (2.16), these have been calculated explicitly for several test spaces and piecewise linear trial spaces by Scotney (1982).

Symmetrization with $B_1(\cdot,\cdot)$

In one dimension with constant a, the best piecewise linear fit in the norm $\|\cdot\|_{B_1}$ is exact at the nodes. Thus it is natural to analyse methods based on finite difference arguments under the $\|\cdot\|_{B_1}$ norm. The Riesz representer R_1 for the model problem (2.16) can be written explicitly as

$$(R_1 w)(x) = w(x) + (b/a)\int_0^x (w(t) - \bar{w})dt, \tag{2.41}$$

where $\bar{w} = \int_0^1 w(t)dt$, which shows its non-local character. For the same problem the earliest upwind test functions were those due to Christie et al. (1976) and Heinrich et al. (1977): if $\phi_i(x)$ are the piecewise linear basis functions, typical of such test functions are

$$\psi_i(x) := \phi_i(x) + \alpha\sigma_i(x) \tag{2.42a}$$

where

$$\sigma_i(x) := \begin{cases} 3(x-x_{i-1})(x_i-x)/(x_i-x_{i-1})^2 & x_{i-1} \leqq x \leqq x_i \\ -3(x_{i+1}-x)(x-x_i)/(x_{i+1}-x_i)^2 & x_1 \leqq x \leqq x_{i+1}. \end{cases} \tag{2.42b}$$

On a uniform mesh, setting the parameter α equal to ξ defined in (2.21) leads to the Allen and Southwell difference operator.

With variable coefficients local values of α are used and, in two dimensions if bilinear elements on rectangles are used, the trial basis functions are the product functions $\phi_i(x)\phi_j(y)$ which are matched with product test functions $\psi_i(x)\psi_j(y)$ with the two parameters α based on the x and y components of $\underline{b}$.

An alternative approach is that due to Hughes and Brooks (1979, 1982): their streamline diffusion method starts from regarding the Allen and Southwell scheme as written in (2.20) as enhancing the diffusion in the direction of the flow vector $\underline{b}$. Thus the scalar diffusion coefficient a of (2.8a) is replaced by the tensor diffusivity with components

$$A_{\ell m} = a\,\delta_{\ell m} + \tilde{a}\,b_\ell\,b_m \tag{2.43a}$$

where
$$\tilde{a} = \tfrac{1}{2}(\xi_1 b_1 h_1 + \xi_2 b_2 h_2) \tag{2.43b}$$

and b_1, b_2 are the components of $\underline{b}$ along the sides of a rectangular element of sides h_1, h_2 : ξ_1, ξ_2 are corresponding values of the parameter (2.21). If this modified operator is used with the Galerkin method and bilinear elements, it can be shown that it is equivalent to using a Petrov-Galerkin method with the test functions

$$\psi = \phi + (\tilde{a}/|b|^2)\underline{b}\cdot\underline{\nabla}\phi. \tag{2.44}$$

These are discontinuous and therefore non-conforming elements. So the terms in the bilinear form corresponding to $(a\underline{\nabla}\phi, \underline{\nabla}\psi)$ have to be evaluated element by element and also the error analysis leading to (2.28) does not strictly apply. Nevertheless evaluation of (2.40) for these two test functions (2.42) and (2.44) does show how effectively these Petrov-Galerkin methods overcome the deficiencies of the pure Galerkin method. The results obtained by Scotney (1982) are given in Table 1.

Table 1

Ratios of Petrov-Galerkin error to optimal error given by $(1-\Delta_1^2)^{\frac{1}{2}}$ - see (2.28) and (2.40)

bh/a	Galerkin	Heinrich et al	Hughes and Brooks
2	1.1547	1.0060	1.0924
5	1.7559	1.0468	1.1509
50	14.468	1.2022	1.1547
500	144.34	1.2344	1.1547
10^5	28868	1.2383	1.1547

The optimal test space for the model problem under $B_1(\cdot,\cdot)$ was used by Hemker (1977), though derived in a different way and with a local basis. The inversion of R_1 to calculate the ψ_i^* of (2.29) gives rise to rather awkward exponentials which are difficult to handle in the formulation (2.30): but Hemker's test functions are quite simple in form (though still difficult to evaluate), namely

$$\psi_i(x) := \begin{cases} [1-e^{-b(x-x_{i-1})/a}]\big/[1-e^{-b(x_i-x_{i-1})/a}], & x_{i-1} \leqq x \leqq x_i \\ [e^{-b(x-x_i)/a} - e^{-b(x_{i+1}-x_i)/a}]\big/[1-e^{-b(x_{i+1}-x_i)/a}], & x_i \leqq x \leqq x_{i+1}. \end{cases} \tag{2.45}$$

These again give rise to the Allen & Southwell difference approximation to -au" + bu' but now sample the right-hand side f of (2.16a) so as to always give exact nodal values. Unfortunately it is much more difficult to extend these test functions into two dimensions and this has not yet been satisfactorily achieved.

Symmetrization with $B_2(\cdot,\cdot)$

With its lack of dependence on b/a it is not clear that $\|\cdot\|_{B_1}$ is an appropriate norm, especially for singular perturbation problems. The alternative bilinear form $B_2(\cdot,\cdot)$ defined in (2.39) has therefore been used by Barrett and Morton (1980, 1981, 1983) in their work. For the model problem, R_2^{-1} now has a simpler form than R_2. Thus the solution of (2.29) can be written explicitly as

$$\psi_i^*(x) = \phi_i(x) + (b/a)\int_x^1 [\phi_i(t) - ce^{-bt/a}]dt, \qquad (2.46)$$

where the constant c is such as to ensure that $\psi_i^*(0) = 0$. Moreover it is easy to use the symmetric form of the equations given by (2.30) which becomes

$$B_2(U_2^*,\phi_i) = (f_2,\phi_i) - cb[u(0) - e^{-b/a}u(1)], \qquad (2.47)$$

where

$$f_2(x) = f(x) + (b/a)[F(x) - \bar{F}], \qquad (2.48)$$

$F(x) = \int_0^x f(t)dt$ and $\bar{F}$ is the average of F with weighting function $e^{-bx/a}$. These formulae are generalised to variable coefficients and Neumann boundary conditions in the above references where examples of their use are given as well as sharp error bounds of the form (2.35): see also Rheinhardt (1982). In two dimensions it is unlikely that R_2^{-1} can be calculated explicitly and various approximate approaches have been tried: several work well for limited classes of problem but at the moment a direct approach to (2.30) appears to be the most generally successful.

A distinctive feature of the optimal piecewise linear approximations in the $\|\cdot\|_{B_2}$ norm obtained by these methods is that steep boundary layers appear as damped oscillations at the mesh frequency. This is because for large Péclet numbers $\|\cdot\|_{B_2}$ tends to the L_2 norm. How sub-gridscale information can be recovered from these results will be discussed in the next section.

3. SUPERCONVERGENCE AND OPTIMAL RECOVERY

As we have seen, one of the main features of a finite element approximation is its optimal, or almost optimal, approximation property in an energy norm, as in (1.8) or (2.28). We then have the problem of recovering from this pointwise estimates of the solution u or its derivatives. Clearly one could use corresponding point values of the approximation U or its derivatives. This is seldom very efficient, for one usually has more a priori knowledge of u than was used in constructing U - such as extra smoothness - and by using this one can achieve much more.

As a simple starting point consider the trivial problem:-

$$-u'' = f \text{ on } (0,1) \text{ with } u(0) = u(1) = 0. \tag{3.1}$$

Let $U(x) = \Sigma U_j\phi_j(x)$ be the piecewise linear Galerkin approximation on a non-uniform mesh with points x_j. Then the Galerkin equations (1.7) reduce to

$$\int_0^1 (u'-U')\phi_j' dx = 0, \text{ i.e. } \frac{\Delta_-\varepsilon_j}{\Delta_-x_j} - \frac{\Delta_-\varepsilon_{j+1}}{\Delta_-x_{j+1}} = 0, \tag{3.2}$$

because ϕ_j' is piecewise constant, where $\varepsilon_j = u(x_j) - U_j$. It follows from the boundary conditions that $\varepsilon_j = 0$ for all j so that U has exact nodal values: note that this requires that the integrals in (f,ϕ_j) are evaluated exactly. To obtain values of u at intermediate points interpolation may be used: linear interpolation just reproduces the corresponding values of U; but if the smoothness of f implies continuity of higher derivatives of u, higher order interpolation using more nodal values will give greater accuracy - or at least a higher order of accuracy. Interpolation theory also indicates how the derivative of u may be recovered to any order of accuracy allowed by its smoothness. Clearly U' itself is only first order accurate at most points: but it will be second order accurate at the mid-point of each interval - the simplest example of the phenomenon of superconvergence.

Another way of looking at the pointwise superconvergence of U, and indeed that which led Hemker to the test functions (2.45), is the following. For the problem (2.1) define the adjoint Green's function $G_{\underline{\xi}}$ by

$$B(v,G_{\underline{\xi}}) = (\delta_{\underline{\xi}},v) \qquad \forall\, v \in H^1_{E_0}, \tag{3.3}$$

where $\delta_{\underline{\xi}}$ is the delta function centred at $\underline{\xi}$. Then if U is the Petrov-Galerkin approximation given by (2.23) and (2.25) we have

$$u(\underline{\xi}) - U(\underline{\xi}) = B(u-U, G_{\xi})$$

$$= B(u-U, G_{\underline{\xi}}-W) \qquad \forall W \in T_0^h. \tag{3.4}$$

Thus from (2.3a) we get,

$$|u(\underline{\xi}) - U(\underline{\xi})| \leq K \|u-U\|_B \|G_{\underline{\xi}}-W\|_B \qquad \forall W \in T_0^h. \tag{3.5}$$

In the case of (3.1), $B(\cdot,\cdot)$ is symmetric and we use the Galerkin method with piecewise linears: the crucial fact is that G_ξ is also piecewise linear, with a change of gradient at ξ. Thus if ξ is a node, G_ξ can be exactly matched from $S_0^h = T_0^h$ and $U(\xi)$ is exact. For the model problem (2.16) the Green's function has exponential form and this led to the choice of test functions (2.45) to obtain exact nodal values.

In two dimensions (or for more complicated problems in one dimension) both of these arguments break down and one cannot achieve exact nodal values: the best piecewise linear fit in the Dirichlet norm (1.9) no longer interpolates at the nodes; and the Green's function is no longer piecewise linear. But much of value can be achieved, especially for the gradients or fluxes, by pointwise sampling.

Superconvergence of gradients for Poisson's equation

For a dozen years or more use has been made of the experimentally observed fact that for bilinear elements over rectangles the gradient has exceptional accuracy at the centroid of each element. Subsequently in a series of papers Zlamal (1977, 1978) and LeSaint and Zlamal (1979) have shown rigorously that the bilinear element is superconvergent at the centroids and that similar higher order elements are superconvergent at corresponding Gauss points: moreover, this is true for more general self-adjoint equations and for mildly non-rectangular quadrilaterals. Meanwhile various corresponding results have been hypothesised for linear elements over triangles but nothing had been established clearly until quite recently. Now in a report which has yet to be published Levine (1983) has clearly set out the true situation. His results provide a good illustration of the recovery problem in a relatively simple situation: the methods of proof that he used are based on those of Zlamal so we begin by outlining these.

Consider the approximation of Poisson's equation using bilinear elements on rectangles of diameter h. Let u be the exact solution and u^I its interpolant by bilinears. Then writing $a(u,w)$ for $(\underline{\nabla} u, \underline{\nabla} w)$ the first result to be established is that, for some constant C,

$$a(u-u^I, W) \leq Ch^2 |u|_{3,\Omega} \, |W|_{1,\Omega} \qquad \forall W \in S_0^h, \tag{3.6}$$

where $|\cdot|_{p,\Omega}$ denotes the pth Sobolev semi-norm over Ω (L_2 norm of all p^{th} order derivatives). This is established through use of the Bramble-Hilbert lemma as follows: define the following linear functional for functions $\tilde{u}$ of the local co-ordinates (ξ,η) over the unit square S on which W is bilinear,

$$F(\tilde{u}) := \int_S \partial_\xi(\tilde{u}-\tilde{u}^I)\partial_\xi \tilde{W} d\xi d\eta; \tag{3.7a}$$

then it is easy to see that

$$|F(\tilde{u})| \leqq C \|\tilde{u}\|_{3,S} \|\partial_\xi \tilde{W}\|_{0,S} \tag{3.7b}$$

where $\|\cdot\|_{p,S}$ denotes the pth Sobolev norm over S; moreover, a little computation enables one to show that for any quadratic polynomial over S we have

$$F(q) = 0; \tag{3.7c}$$

it is then a direct result of the lemma that

$$|F(\tilde{u})| \leqq C |\tilde{u}|_{3,S} \|\partial_\xi \tilde{W}\|_{0,S}. \tag{3.7d}$$

The required result (3.6) can then be obtained by scaling and summing over all rectangles. We next use this result with $W = U - u^I$, where U is the Galerkin approximation satisfying (1.7), to obtain

$$|U-u^I|^2_{1,\Omega} = a(U-u^I,\ U-u^I) = a(u-u^I,\ U-u^I)$$

$$\leqq Ch^2|u|_{3,\Omega}\ |U-u^I|_{1,\Omega}$$

$$\text{i.e. } |U-u^I|_{1,\Omega} \leqq Ch^2|u|_{3,\Omega}\ . \tag{3.8}$$

Thus U is an order of magnitude closer to u^I than it is to u, in the energy norm.

Suppose now that D_P is a sampling operator at a point P, for instance for the derivative ∂_x at the centroid of a rectangular element R. Then by a similar argument to the above, using the Bramble-Hilbert lemma and a computation for quadratic polynomials, one can show that

$$|D_P(u-u^I)| \leqq Ch|u|_{3,R} \ . \tag{3.9}$$

Finally, by writing $D_P(u-U) = D_P(u-u^I) + D_P(u^I-U)$, using both (3.8) and (3.9) and summing over all rectangles R_j in Ω we obtain

$$[\Sigma_{(j)} h^2 \ |D_{P_j}(u-U)|^2]^{\frac{1}{2}} \leqq Ch^2|u|_{3,\Omega} \ . \tag{3.10}$$

This holds for ∂_x at points on the vertical bisector of each rectangle and for ∂_y at points on the horizontal bisectors: hence it holds for the gradient at the centroids and so confirms the superconvergence phenomenon in this ℓ_2 sense.

For linear elements over triangles, Levine (1983) has shown both theoretically and numerically that there are no points where the gradient is superconvergent. However, suppose the triangulation is such that there are six triangles per vertex and the triangles can be grouped in pairs to form parallelograms with vertical diagonals and also in pairs to form parallelograms with horizontal diagonals. Then he has proved the conjecture of Strang and Fix (1973, p. 169) that the derivatives along the edges of the triangles are superconvergent at the mid-points. Moreover, he has shown that the average of the normal derivative in the two triangles either side of an edge is also superconvergent at the mid-point. Thus the gradient can be recovered to second order accuracy at the edge mid-points by this very simple device: averaging between the three mid-points of a triangle will also give the gradient to second order accuracy at the centroids. The proofs of these results follow similar lines to those of Zlamal, outlined above, but more constructive methods than the use of the Bramble-Hilbert lemma give sharper bounds for several of the results. Numerical experiments confirm the practical value of the results and the importance of the triangulation giving six triangles per node. The regularity of the mesh can otherwise be considerably relaxed and there is some hope that similar results can be proved in the supremum norm.

It is interesting to note that the recovery procedures described above, followed by interpolation, will often coincide with the use of divided differences of the Galerkin approximation as advocated, for instance, by Long and Morton (1977) and Thomée (1977). The analysis of the first reference, however, though also covering quadratic elements was essentially limited to regular meshes: that in the latter did not cover linear elements. Finally, before leaving this topic we should note complementary results of Douglas, Dupont and Wheeler (1974) and Wheeler (1974) for recovering the normal gradient at a Dirichlet boundary, which is often of very great practical importance: indeed, the procedure whose superconvergence is established in this reference was proposed for calculating boundary fluxes in heat-transfer problems by Wheeler (1973).

Optimal recovery

In their seminal paper, Golumb and Weinberger (1959) explored many of the basic ideas of optimal recovery which are pertinent to finite element methods: see also Micchelli and Rivlin (1976) for a more recent survey. The general situation is as follows: we are given the values of n linear data functionals $F_1(u), F_2(u), \ldots, F_n(u)$ of an unknown function u together with some (non-linear) constraint on u, such as $|u|_p \leqq K$; then the problem is to define an optimal estimator for another linear functional F(u), that is one with a minimal a priori error bound. As applied to finite element methods, the ideas are related to that of the hypercircle (Synge, 1955). For consider the problem (1.2), but with homogeneous boundary data, for $u \in H^1_{E_0}$: the Galerkin approximation U of (1.5a), (1.6) is determined from the data functionals

$$F_i(u) := A(u,\phi_i) = \langle f,\phi_i\rangle \qquad \forall\phi_i \in S^h_0; \tag{3.11}$$

and it is easy to check that U coincides with the centre $\bar{u}$ of Synge's hypercircle defined by

$$\|\bar{u}\|_A = \inf\{\|v\|_A \mid v\in H^1_{E_0} \text{ s.t. } A(v,\phi_i) = (f,\phi_i)\ \forall\phi_i \in S^h_0\}. \tag{3.12}$$

Now suppose F(u) is to be estimated and define $\bar{y}$ by

$$F(\bar{y}) = \sup\{|F(v)|\,|v \in H^1_{E_0} \text{ s.t. } \|v\|_A = 1,\ A(v,\phi_i) = 0\quad \forall\phi_i \in S^h_0\}. \tag{3.13}$$

Then F(U) is an optimal estimator of F(u) with

$$|F(u) - F(U)|^2 \leqq |F(\bar{y})|^2\,[\|u\|^2_A - \|U\|^2_A], \tag{3.14}$$

given that the constraint on u is of the form $\|u\|_A \leqq r$. This bound is sharp with F(U) lying at the centre of the range of possible values for F(u) obtained by taking $u = U + \alpha(r^2 - \|U\|^2_A)\bar{y}$ with $|\alpha| \leqq 1$. It is important to note that although $\bar{y}$ depends on F, U is quite independent of the linear functional to be estimated.

For example, consider the one-dimensional self-adjoint problem

$$\begin{aligned} -(pu')' + qu &= f \qquad \text{on } (a,b) \\ \text{with}\qquad u(a) = u(b) &= 0 \end{aligned} \tag{3.15}$$

and $p > 0$, $q \geqq 0$. Suppose we wish to estimate $u(\xi) =: F(u)$ for $\xi \in (a,b)$. Then it is easy to check that $\bar{y}$ is the difference $G(x,\xi) - G^h(x,\xi)$ of the Green's function from its best fit from S_0^h. We then obtain

$$|u(\xi) - U(\xi)|^2 \leqq [G(\xi,\xi) - G^h(\xi,\xi)][\|u\|_A^2 - \|U\|_A^2], \quad (3.16)$$

which is actually the same as (3.5) in this case. When the trial space is piecewise linear, one can deduce that the nodal values are optimal sampling points (in the sense of giving a minimal error bound) although in general one will still obtain only first order accuracy unless stronger smoothness hypotheses are made on u than merely the boundedness of $\|u\|_A$: this indicates how singular was the situation of $p = 1$, $q = 0$ covered by (3.2).

This last point also shows the limitations of this framework: for it is not clear how one can exploit any greater smoothness, that u is known to possess, in the estimation of $F(u)$; we shall take this up in the next sub-section on defect correction. In the meantime let us consider the other extreme, appropriate to the diffusion-convection problems of section 2 and the hyperbolic problems of the next section, where the solution u may be far from smooth and typified by $|q| \gg |p|$ in (3.15). The Galerkin approximation to (3.15) gives the best fit from the trial space $S_0^h = \text{span}\,\{\phi_j\}$ in the energy norm

$$\left\{\int_a^b (pu'^2 + qu^2)\,dx\right\}^{\frac{1}{2}}; \quad (3.17)$$

in the limit $p \to 0$ this becomes the weighted L_2 norm. For the recovery problem in this limit there are two natural sets of data functionals, the moment functionals

$$F_i^M(u) := \int_a^b qu\phi_i\,dx \quad (3.18a)$$

and the point functionals given by the nodal parameters of the best fit

$$F_i^P(u) := U_i. \quad (3.18b)$$

Barrett, Moore and Morton (1983) consider the <u>local recovery problem</u> for both types: that is, the problem corresponding to classical interpolation of approximating u from n consecutive values of either (3.18a) or (3.18b). They show that for piecewise linear ϕ_i there exists a unique $(n-1)^{th}$ degree polynomial P_{n-1} with the same set of data functional values and that if $u \in H^n(I)$ then

$$|u(x) - P_{n-1}(x)| \leqq Ch^n |u|_{n,I}, \qquad x \in I, \tag{3.19}$$

where h is the maximal node spacing and I is the union of the support of the basis functions ϕ_i which are involved. Moreover, if $u \in H^{n+1}(I)$ there are n points of superconvergence in I where an extra order of accuracy is achieved. These results generalise and extend those given by Bramble and Schatz (1977). As a particular case they include the following well-known result: for n = 3, q = 1 and on a uniform mesh, the parabola which has the point functionals U_{j-1}, U_j and U_{j+1} yields the superconvergent recovery result

$$|u(x_j) - \frac{1}{12}[U_{j-1} + 10U_j + U_{j+1}]| \leq \text{const. } h^4 |u^{(iv)}|. \tag{3.20}$$

Such local recovery formulae are of great practical value in extracting the best results from Galerkin approximations and have been used by a number of authors. They hold also for $p \neq 0$ so long as $p/q = O(h^2)$: thus they apply to the $\|\cdot\|_{B_2}$ norm of (2.39a) used by Barrett and Morton (1980) in the diffusion-convection problem, for any moderately large mesh Peclet number; these authors also used special formulae based on exponentials rather than polynomials for recovery of boundary layers much narrower than the mesh spacing.

Defect correction

For general p,q in (3.15), however, local recovery is not possible: and direct global recovery would often be prohibitively expensive. So consider the following approach based on assuming that $u \in H^2(a,b)$. Introduce the bilinear form

$$\tilde{A}(v,w) = \int_a^b (pv''w'' + qv'w')dx \tag{3.21}$$

and the representers χ_i of the data functionals F_i^M in this form:-

$$F_i^M(u) := \int_a^b (pu'\phi_i' + qu\phi_i)dx \equiv A(u,\phi_i) = \tilde{A}(u,\chi_i), \tag{3.22}$$

where we will take the ϕ_i as the piecewise linears. Then applying the hypercircle result to this form, the centre of the hypercircle for the optimal estimation problem in which the $F_i^M(u)$ are given is an element $\tilde{U}$ in the span of these representers given by

$$A(u - \tilde{U}, \phi_i) = 0 \qquad \forall i, \tag{3.23a}$$

$$\text{i.e.} \quad \tilde{A}(u - \tilde{U}, \chi_i) = 0 \qquad \forall i. \tag{3.23b}$$

That is, we have a Petrov-Galerkin approximation in the original bilinear form $A(\cdot,\cdot)$ which is also a Galerkin approximation in the new form $\tilde{A}(\cdot,\cdot)$. When p and q are constants it is clear that $\chi_i'' = -\phi_i$ and hence that the χ_i are natural cubic splines: thus from an error analysis of either (3.23b) or of (3.23a) (using the theorem of (2.27) and (2.28)) we find that $\tilde{U}$ achieves fourth order accuracy when $u \in H^4(a,b)$.

Now solving either of the forms (3.23) directly is a completely separate computation from calculating the piecewise linear Galerkin approximation U. So instead we consider the calculation of $\tilde{U}$ as a recovery operation: it is more convenient to use cubic splines even for variable p,q so we define a natural cubic spline U^c by

$$\begin{aligned} A(U - U^c, \phi_i) &= A(u - U^c, \phi_i) = 0 \qquad \forall i \\ U^c(a) &= U^c(b) = 0. \end{aligned} \tag{3.24}$$

Let P be the nodal interpolatory mapping from piecewise linears to natural cubic splines. Then we define the iteration for piecewise linears $U^{(\ell)}$ by

$$\begin{aligned} A(U^{(\ell+1)}, \phi_i) &= A(U^{(\ell)}, \phi_i) - A(PU^{(\ell)} - U, \phi_i) \qquad \forall i \\ U^{(\ell+1)}(a) &= U^{(\ell+1)}(b) = 0, \end{aligned} \tag{3.25}$$

with $U^{(0)} = U$. The last term of (3.25) can be rewritten as

$$A(PU^{(\ell)}, \phi_i) - (f, \phi_i)$$

which reveals its rôle as a 'defect' in the calculation of the Petrov-Galerkin approximation U^c to u and (3.25) as a defect correction technique. In Barrett, Moore and Morton (1983) it is shown that the convergence factor is $O(h)$ if $p' \in L^\infty(a,b)$ and $O(h^2)$ if p is constant. Thus one or two iterations, which we note involve only the original stiffness matrix $A(\phi_j, \phi_i)$ for the piecewise linears, are sufficient to obtain the full fourth order accuracy of U^c.

4. HYPERBOLIC PROBLEMS

At first sight hyperbolic problems offer an unpromising field for the use of finite element methods and, indeed, finite difference methods continue their domination in practical problems with perhaps the strongest challenge at the moment coming from spectral methods. This inauspicious prospect is because of a lack of useful variational principles and the fact that many of the phenomena are local in character and less suitable for global approximation. To accentuate the difference from elliptic problems we shall exclude from our consideration the sort of steady hyperbolic problems that occur in supersonic gas flow. Thus we shall assume that the time t is one independent variable and consider first order systems of the form

$$\underset{\sim}{u}_t + L(\underset{\sim}{u}) = 0, \tag{4.1}$$

where $\underset{\sim}{u} = \underset{\sim}{u}(\underline{x},t)$ is a vector of unknowns, the subscript t or the operator ∂_t denotes partial differentiation with respect to t and L is a (generally non-linear) operator involving the first order spatial derivatives.

Then the first choice is whether to use finite element approximation in time as well as space. We shall not do so but use finite differences in the time variable. This is partly for simplicity and flexibility: but it is mainly because with finite elements we seek approximations which are optimal in an integral norm and this would seem to be more appropriate in just the space variables rather than in space-time. We shall moreover concentrate on <u>one-step methods</u> in time and indeed mainly on the explicit Euler method, with which much can be achieved: the methods derived can then be extended to implicit methods or, by predictor-corrector or Runge-Kutta schemes, to higher order methods. We shall say little about boundary conditions in this section.

Petrov Galerkin methods

These have been widely used to overcome the disadvantages of pure Galerkin methods, just as with non-self-adjoint elliptic problems. Thus at time-step n we write the approximation as

$$\underset{\sim}{U}^n(\underline{x}) = \Sigma_{(j)} \underset{\sim}{U}_j^n \phi_j(\underline{x}) \tag{4.2}$$

in terms of trial space basis functions $\phi_j(\underline{x})$. Then the Petrov-Galerkin method for (4.1) based on Euler time-stepping and test space basis functions $\psi_i(\underline{x})$ has the form

$$\left(\frac{\underset{\sim}{U}^{n+1} - \underset{\sim}{U}^n}{\Delta t} + L(U^n),\ \psi_i \underset{\sim}{e}_{(k)} \right) = 0 \qquad \forall i,\ k \tag{4.3}$$

where the vector $\underset{\sim}{e}_{(k)}$ has a single unit component in k^{th} position. The Galerkin method, with ψ_i taken as ϕ_i, has advantages for small Δt:

if $L(\cdot)$ is a conservative operator, that is $(L(\underset{\sim}{u}),\underset{\sim}{u}) = 0$ so that the L_2 energy $\|\underset{\sim}{u}\|^2$ is conserved, this same property is retained for $\|\underset{\sim}{U}\|^2$; also as $\Delta t \to 0$, Galerkin equations can be regarded as giving the best L_2 fit to $\partial_t \underset{\sim}{U}$ when $L(\underset{\sim}{U}^n)$ is known. However Galerkin methods generally have very poor stability properties as well as poor accuracy for moderate values of Δt. For example, for the linear advection equation with $L(\underset{\sim}{u})$ replaced by $a\partial u \partial/x$, just as for (2.16), (2.18) we obtain a central difference approximation scheme and with explicit time-stepping this is stable only for $\Delta t = O(h^2)$.

The linear advection problem is a natural model problem for the development of more effective test functions and most of those described in the literature owe something to the idea of upwinding. On a uniform mesh the key parameter is $a\Delta t/h$, the CFL number (after Courant, Friedrichs and Lewy): if this has an integral value, clearly the solution can be exactly advected on the mesh. This is a particularly pertinent property within the framework that we have taken, and is usually satisfied by difference methods but not by Galerkin methods. Thus the unit CFL property was taken as a specific objective by Morton and Parrott (1980) in devising a variety of Petrov-Galerkin methods. It would be inappropriate to describe their results in detail here and we give merely a flavour. For linear ϕ_j in one dimension and various time-stepping schemes they found special test functions χ_i with the unit CFL property and then for scalar problems set

$$\psi_i(x) = (1 - \nu)\phi_i(x) + \nu\chi_i(x) \tag{4.4}$$

where ν is determined from the CFL number. With the Euler scheme this gives a method close to but more accurate than the well-known Lax-Wendroff difference scheme: with Crank-Nicolson it gives a third order accurate scheme and with leapfrog one of fourth order. The methods also extend to systems of equations.

However, Morton and Stokes (1982) found that some properties were difficult to extend into two space dimensions. While the unit CFL property could be retained with bilinear elements on rectangles it could not be made to hold along all the edge directions of a uniform triangular mesh when piecewise linear elements were used. Also the test functions χ_i are discontinuous and in the case of the Euler scheme do not span the unit function: thus even the conservation of the first moment $\int u\,dx$ is lost. The attention of this author has therefore shifted to the Characteristic Galerkin methods described below.

Euler Characteristic Galerkin (ECG) methods

These make more explicit use of characteristics. Consider the scalar conservation law in one space dimension (on the whole real line)

$$\partial_t u + \partial_x f(u) = 0 \tag{4.5a}$$

or

$$\partial_t u + a(u)\partial_x u = 0 \tag{4.5b}$$

where $a(u) = \partial f/\partial u$. Then u is constant along the characteristics $dx/dt = a$ so that if we write $u^n(x)$ for $u(x,n\Delta t)$ and use a similar notation for f and a, we have for smooth flows

$$u^{n+1}(y) = u^n(x) \text{ where } y = x + a^n(x)\Delta t. \tag{4.6}$$

Thus the L_2 projection of u^{n+1} onto the trial space S^h spanned by $\{\phi_j\}$ is related to that of u^n by

$$(u^{n+1}-u^n,\phi_i) = \int_{-\infty}^{\infty} u^{n+1}(y)\phi_i(y)dy - \int_{-\infty}^{\infty} u^n(x)\phi_i(x)dx$$

$$= \int_{-\infty}^{\infty} u^n(x)[\phi_i(y)\frac{dy}{dx} - \phi_i(x)]dx$$

$$= \int_{-\infty}^{\infty} u^n(x)\ [\frac{d}{dx}\int_x^y \phi_i(z)dz]dx$$

$$=-\int_{-\infty}^{\infty} \partial_x u^n(x)\ [\int_x^y \phi_i(z)dz]dx:$$

that is, we have the exact relationship for the true solution

$$(u^{n+1} - u^n,\phi_i) + \Delta t(\partial_x f^n,\Phi_i^n) = 0 \qquad \forall\phi_i \in S^h \tag{4.7}$$

where

$$\Phi_i^n(x) := \frac{1}{a^n(x)\Delta t}\int_x^{x+a^n(x)\Delta t} \phi_i(z)dz\ . \tag{4.8}$$

This form strongly suggests the following basic ECG method:-

$$(U^{n+1} - U^n,\phi_i) + \Delta t(\partial_x f(U^n),\ \bar{\Phi}_i^n) = 0 \qquad \forall\phi_i \in S^h, \tag{4.9}$$

where $\bar{\Phi}_i^n$ has the same form as Φ_i^n in (4.8) but with a^n replaced by $a(U^n)$. This method in effect exactly traces the evolution of $U^n(x)$ through one time-step, as given by the relationship (4.6), and then projects this onto S^h. However, it is implemented like a Petrov-Galerkin method, from which it is distinguished by the fact that the time difference involves a Galerkin inner product (and hence only symmetric equations have to be solved) while the spatial operator is combined with a test function directly derived from the trial function. The fact that the derivation of $\bar{\Phi}_i^n$ involves only an averaging process means that it may be very efficiently approximated. Several approximations when the ϕ_j are piecewise linear are given in Morton (1982b) for the CFL number range $|a\Delta t/h| \leq 1$: these reproduce the results of (4.9) when a is constant and one of them involves evaluation of only the same inner products used in the Galerkin method if the <u>product approximation scheme</u> is used for $\partial_x f$, namely

$$\partial_x f(U^n) \cong \Sigma_{(j)} f(U_j^n)\phi_j' \; . \tag{4.10}$$

In the same reference, it is also shown how (4.9) extends naturally into more space dimensions: one then has a flux vector $\underline{f}(u)$ and a characteristic velocity vector $\underline{a}(u) = \partial \underline{f}/\partial u$ so that (4.8) is replaced by

$$\phi_i^n(\underline{x}) = \frac{1}{|\underline{a}^n(\underline{x})|\Delta t} \int_{\underline{x}}^{\underline{x}+\underline{a}^n(\underline{x})\Delta t} \phi_i(\underline{z})\,d\ell, \tag{4.11}$$

integration being along the straight line between $\underline{x}$ and $\underline{x} + \underline{a}^n(\underline{x})\Delta t$.

It should be noted that in principle there is no stability limit for (4.9). Indeed, since if the terms in (4.9) are evaluated exactly the only error is at the projection stage, the least error is committed in going from $t = 0$ to $t = T$ if one large step $\Delta t = T$ is used! This is not very practical of course because for a system of equations the characteristics will be curved, the simple relation (4.6) will not hold and shocks will often intervene to destroy the basic assumption above that the solution is smooth. However, for conventional time-steps, with CFL numbers of the order of unity, these methods are extremely accurate, piecewise linear elements giving third order accuracy for instance. And when they are written out in terms of nodal values they show many relationships with high accuracy difference methods. These relationships and a detailed analysis of the basic ECG scheme will be given in Morton (1984). Here we point out just one such set of links. Suppose that in deriving (4.7) and (4.9) we used a mixed norm instead of the L_2 norm, i.e. the inner product

$$(u,v) + \gamma^2(\partial_x u, \partial_x v) \tag{4.12}$$

for some constant γ. Then, as for (4.8) with piecewise linear ϕ_j on a uniform mesh and $a\Delta t/h \in (0,1)$, the corresponding special test function has support over three intervals (x_{i-2}, x_{i+1}): however, if $\gamma^2 = \frac{1}{6}(a\Delta t)^2$ its average value over (x_{i-2}, x_{i-1}) is zero and it can therefore be approximated by $\phi_i + \frac{1}{2}a\Delta t\phi_i'$, which gives the same scheme if a is constant. The resulting scheme is then one of the Taylor-Galerkin schemes given by Donea (1983), which were derived in an entirely different manner, and in turn is equivalent to the EPG II scheme given by Morton and Parrott (1980) which was derived as indicated above in (4.3) and (4.4). Indeed, all of Donea's schemes can be derived in a like manner from the Characteristic-Galerkin methods.

Much more yet can be derived from the basic formulae (4.7) and (4.9). If U^n is the best fit to u^n from S^h in either the L_2 norm or the mixed norm derived from (4.12) any further knowledge of u^n (derived for example from studies of the original differential equation (4.5)) can be exploited through the recovery techniques discussed in section 3. Thus suppose this further information is incorporated in a recovery function $\tilde{u}^n$ which in the L_2 case satisfies

$$(\tilde{u}^n - U^n, \phi_i) = 0 \qquad \forall \phi_i \in S^h . \tag{4.13}$$

Then (4.9) can be replaced by

$$(U^{n+1} - U^n, \phi_i) + \Delta t(\partial_x f(\tilde{u}^n), \tilde{\Phi}_i^n) = 0 \qquad \forall \phi_i \in S^h , \tag{4.14}$$

where $\tilde{\Phi}_i^n$ has the same form as Φ_i^n in (4.8) but with a^n replaced by $a(\tilde{u}^n)$. For example, if u^n is smooth enough one can as in section 3, cf. (3.24), recover from piecewise linears to cubic splines. More interestingly, this enables one to use the non-conforming piecewise constant basis functions: as is shown in Morton (1984) for the linear advection problem with constant a, quadratic spline recovery from piecewise constants yields through (4.14) a formula identical to (4.9) with piecewise linears. There is in fact a whole hierarchy of similarly related Characteristic Galerkin methods based on splines.

Piecewise constant ϕ_i are a natural choice for shock-modelling and their use in connection with (4.13) and (4.14) has been explored in (Morton, 1982c). Clearly the basic ECG scheme (4.9) is not defined when U^n, $f(U^n)$ and $a(U^n)$ all have discontinuities at the cell boundaries which we can take to be at $x_{i+\frac{1}{2}}$. This is true even for smooth flows: but we can then spread the discontinuities by a linear variation over $\frac{1}{2}\theta h$ either side of $x_{i+\frac{1}{2}}$ to join constant values $\tilde{u}_i$ and $\tilde{u}_{i+1}$ either side; then it is easy to see that on a uniform mesh (4.13) gives

$$\tilde{u}_i + \frac{\theta}{8}\,\delta^2\tilde{u}_i = U_i \qquad \forall i. \tag{4.15}$$

For sufficiently small θ and if $a(\tilde{u}_{i-1})$, $a(\tilde{u}_i)$ and $a(\tilde{u}_{i+1})$ are all non-negative, we also find that (4.14) reduces to

$$h(U_i^{n+1} - U_i^n) + \Delta t\,[\Delta_- f(\tilde{u}_i) + \frac{\theta}{8}\frac{h}{\Delta t}\,\delta^2\tilde{u}_i] = 0 \qquad \forall i \tag{4.16a}$$

i.e. $$U_i^{n+1} = \tilde{u}_i^n - (\Delta t/h)\Delta_- f(\tilde{u}_i) \qquad \forall i. \tag{4.16b}$$

Clearly as $\theta \to 0$ it reduces to the familiar first order upwind scheme: for $\theta > 0$ it has a similar form in (4.16b) but from (4.16a) it is seen to incorporate an anti-diffusive flux, as in many modern difference schemes; in fact it can be seen from (4.15) that it is the recovery process that is sharpening up the profiles broadened by the averaging process which is presumed to have led to U_i.

Regions of smooth flow are recognised by characteristics not crossing, typically that is $a(\tilde{u}_{i-1}) \leq a(\tilde{u}_i)$. On the other hand $a(\tilde{u}_{i-1}) > a(\tilde{u}_i)$ will lead to crossing of characteristics and the breakdown of (4.6) because the mapping from y to x is not unique. Even if recovery by a smooth function were appropriate in this case, (4.14) would not describe the exact evolution of $\tilde{u}^n$ followed by projection: instead it gives the projection of the multi-valued solution produced by the crossing characteristics from (4.6). The resulting approximation was used successfully by Morton (1982c) to model breaking waves but of course now Δt must be limited if good accuracy is to be achieved. In fact it turns out that the same upwind formula is obtained in the limit $\theta \to 0$ where $a(\tilde{u}_{i-1}) \leq a(\tilde{u}_i)$ or $a(\tilde{u}_{i-1}) > a(\tilde{u}_i)$. Moreover, if $f(\cdot)$ is convex with a single sonic point $\bar{u}$ at which $a(\bar{u}) = 0$, the intermediate case in which $a(\tilde{u}_{i-1})a(\tilde{u}_i) < 0$ is dealt with very naturally through the recovery process: $f(\tilde{u}_i) - f(\tilde{u}_{i-1})$ is split into "waves" $f(\tilde{u}_i) - f(\bar{u})$ and $f(\bar{u}) - f(\tilde{u}_{i-1})$ with the right wave contributing to the updating of U_i^n and the left to that of U_{i-1}^n; the scheme is then identical with that of Engquist and Osher (1980).

It would not be appropriate to go into greater details on shock modelling with the ECG scheme here: the scalar problem is dealt with fully in Morton (1982c) and the use of the approximate Riemann solvers of Roe (1981) to deal with the Euler equations of gas dynamics is described in Morton (1983). The important point here is to recognise the crucial role played by the recovery process and the exploitation of the best L_2 fit property of the approximation in the understanding and development of the methods.

Thus these shock problems represent the furthest point in a consistent line of development that we have presented, starting from the self-adjoint elliptic problems in section 1. There the solutions were smooth and the appropriate norms dominated by derivative terms: here the solutions are discontinuous (and the analysis should ideally be in L_1) and we have used mainly the L_2 norm. But the objective of optimal approximation is consistent throughout.

REFERENCES

Allen, D. and Southwell, R. (1955) Relaxation methods applied to determine the motion, in two dimensions, of a viscous fluid past a fixed cylinder. *Quart. J. Mech. and Appl. Math. VIII*, 129-145.

Babuska, I. and Aziz, A.K. (1972) Survey lectures on the mathematical foundations of the finite element method. "The Mathematical Foundations of the Finite Element Method with Applications to Partial Differential Equations, (A.K. Aziz, Ed.), Academic Press, New York, 3-363.

Barrett, J.W., Moore, G. and Morton, K.W. (1983) Optimal recovery and defect correction in the finite element method. Univ. of Reading, Num. Anal. Report 11/83.

Barrett, J.W. and Morton, K.W. (1980) Optimal finite element solutions to diffusion-convection problems in one dimension. *Int. J. Num. Meth. Engng.*, **15**, 1457-1474.

Barrett, J.W. and Morton, K.W. (1981) Optimal Petrov-Galerkin methods through approximate symmetrization. *IMA J. Numer. Anal.*, **1**, 439-468.

Barrett, J.W. and Morton, K.W. (1983) Approximate symmetrization and Petrov-Galerkin methods for diffusion-convection problems. *Comp. Meths. in Appl. Mech. Engng.* (To appear).

Bramble, J.H. and Schatz, A.H. (1977) Higher order local accuracy by averaging in the finite element method. *Math. Comp.*, Vol. **31**, 94-111.

Christie, I., Griffiths, D.F., Mitchell, A.R. and Zienkiewicz, O.C. (1976) Finite element methods for second order differential equations with significant first derivatives. *Int. J. Num. Meth. Engng.*, **10**, 1389-1396.

Ciarlet, P.G. (1978) The Finite Element Method for Elliptic Problems. North-Holland, Amsterdam.

Donea, J. (1983) A Taylor-Galerkin method for convective transport problems. *Int. J. Num. Meth. in Engng.* (To appear).

Douglas, J. Jr., Dupont, T. and Wheeler, M.F. (1974) A Galerkin procedure for approximating the flux on the boundary for elliptic and parabolic boundary value problems. R.A.I.R.O. R-2, 47-59.

Engquist, B. and Osher, S. (1980) Stable and entropy satisfying approximations for transonic flow calculations. *Math. Comp.*, **34**, 45-75.

Golumb, M. and Weinberger, H.F. (1959) Optimal approximation and error bounds. Symp. on Numerical Approximation (R.E. Langer, Ed.), Madison, 117-190.

Heinrich, J.C., Huyakorn, P.S., Mitchell, A.R. and Zienkiewicz, O.C. (1977) An upwind finite element scheme for two-dimensional convective transport equations. *Int. J. Num. Meth. Engng.*, **11**, 131-143.

Hemker, P.W. (1977) A numerical study of stiff two-point boundary problems. Thesis, Mathematisch Centrum, Amsterdam.

Hughes, T.J.R. (1979) Finite Element Methods for Convection Dominated Flows, AMD Vol. 34, Am. Soc. of Mech. Eng. (New York).

Hughes, T.J.R. and Brooks, A.N. (1979) A multi dimensional upwind scheme with no crosswind diffusion. "Finite Element Methods for Convection Dominated Flows",(T.J.R. Hughes, Ed.), AMD Vol. 34, Am. Soc. Mech. Eng., (New York), 19-35.

Hughes, T.J.R. and Brooks, A.N. (1982) A theoretical framework for Petrov-Galerkin methods with discontinuous weighting functions: application to the streamline-upwind procedure. "Finite Elements in Fluids", Vol. 4, (R.H. Gallagher, D.H. Norrie, J.T. Oden and O.C. Zienkiewicz, Eds.), J. Wiley and Sons, New York, 47-65.

Lesaint, P. and Zlamal, M. (1979) Superconvergence of the gradient of finite element solutions. R.A.I.R.O., *Numer. Anal.*, **13**, 139-166.

Levine, N. (1983) Superconvergent recovery of the gradient from finite element approximation on linear triangles. Univ. of Reading, Num. Anal. Report 6/83.

Long, M.J. and Morton, K.W. (1976) The use of divided differences in finite element calculations. *J. Inst. Maths. Applics.*, **19**, 307-323.

Micchelli, C.A. and Rivlin, T.J. (1976) A survey of optimal recovery. "Optimal Estimation in Approximation Theory",(C.A. Micchelli and T.J. Rivlin, Eds.), Plenum Press, New York, 1-54.

Morton, K.W. (1982a) Finite element methods for non-self-adjoint problems. Proc. SERC Summer School, 1981 (P.R. Turner, Ed.), Lect. Notes in Maths. 965, Springer-Verlag, Berlin, 113-148.

Morton, K.W. (1982b) Generalised Galerkin methods for steady and unsteady problems. Proc. IMA Conf. Num. Meth. for Fluid Dynamics, (K.W. Morton and M.J. Baines, Eds.), Academic Press, 1-32.

Morton, K.W. (1982c) Shock capturing, fitting and recovery. Proc. 8th Int. Conf. on Numerical Methods in Fluid Dynamics, Aachen. (E. Krause, Ed.), Lect. Notes in Physics 170, Springer-Verlag, Berlin, 77-93.

Morton, K.W. (1983) Characteristic Galerkin methods for hyperbolic problems. Proc. 5th GAMM Conf. on Numerical Methods in Fluid Mechanics. (To appear).

Morton, K.W. (1984) Analysis of Characteristic Galerkin methods for scalar problems in one dimension. (in preparation).

Morton, K.W. and Parrott, A.K. (1980) Generalised Galerkin methods for first order hyperbolic equations. *J. Comp. Phys.*, **36**, 249-270.

Morton, K.W. and Stokes, A. (1982) Generalised Galerkin methods for hyperbolic equations. Proc. MAFELAP 1981 Conf. (J.R. Whiteman, Ed.), Academic Press, London, 421-431.

Rheinhardt, H.J. (1982) A-posteriori error analysis and adaptive finite element methods for singularly perturbed convection-diffusion equations. *Math. Methods Appl. Sci.* (To appear).

Roe, P.L. (1981) Approximate Riemann solvers, parameter vectors and difference schemes. *J. Comp. Phys.*, **43**, 357-372.

Scotney, B.W. (1982) Error analysis and numerical experiments for Petrov-Galerkin methods. Univ. of Reading Num. Anal. Report 11/82.

Strang, G. and Fix, G.J. (1973) "An Analysis of the Finite Element Method". Prentice-Hall, New York, 169.

Synge, J.L. (1957) "The Hypercircle in Mathematical Physics". Cambridge University Press, London.

Thomée, V. (1977) High order local approximations to derivatives in the finite element method. *Math. Comp.*, **31**, 652-660.

Wheeler, J.A. (1973) Simulation of heat transfer from a warm pipeline buried in permafrost. Presented on the 74th National Meeting of Am. Inst. of Chem. Eng., New Orleans.

Wheeler, M.F. (1974) A Galerkin procedure for estimating the flux for two-point boundary value problems. *SIAM J. Numer. Anal.*, **11**, 764-768.

Zlamal, M. (1977) Some superconvergence results in the finite element method. "Mathematical Aspects of Finite Element Methods", Springer-Verlag, 353-362.

Zlamal, M. (1978) Superconvergence and reduced integration in the finite element method. *Math. Comp.*, **32**, 663-685.

MIXED FINITE ELEMENT METHODS

P.A. Raviart

(Université Pierre et Marie Curie, Paris, France)

1. INTRODUCTION

Conforming finite element methods (F.E.M.) are now widely popular in the scientific and technical community and are the basis for a number of efficient well documented finite element codes which provide the numerical solution of a large number of problems arising in practice. These conforming F.E.M. are based on classical variational principles or classical variational formulations such as, in linear elasticity, the minimum potential energy principle or equivalently the virtual work principle. However, for some classes of problems, it may be essential to use F.E.M. based on more sophisticated variational principles such as, in elasticity, the complementary energy principle or the Hellinger-Reissner principle. Such variational principles have been first advocated by Fraeijs de Veubeke and Pian in order to construct new F.E.M.: the dual, mixed and hybrid F.E.M.

The purpose of this chapter is to provide a mathematical introduction to such methods. On the one hand, we shall give the main lines of the abstract theory of approximation of problems based on a general two-field variational formulation. On the other hand, in order to illustrate the abstract results, we shall concentrate on mixed F.E.M. and their applications to the numerical solution of problems arising in incompressible fluid mechanics. In particular, we shall give a fairly detailed discussion of the finite element approximation of the Stokes equations. Hence, this chapter may be also considered as an introduction to this problem of great practical importance since it is an essential step in the numerical solution of the incompressible Navier-Stokes equations.

For applications of the abstract theory to mixed hybrid F.E.M., we refer for instance to Raviart and Thomas (1979) in the case of second order elliptic problems and to Brezzi (1979) in the case of the fourth order problems.

2. EXAMPLES

We begin by considering two simple significant model problems arising in fluid mechanics in order to introduce general two-field variational formulations.

2.1 The Dirichlet problem for the Laplace operator

We consider an incompressible fluid flow in a simply connected plane domain Ω with boundary Γ. We denote by $\underline{u} = (u_1,u_2)$ the velocity of the fluid and by

$$\omega = \text{curl } \underline{u} = \frac{\partial u_2}{\partial x_1} - \frac{\partial u_1}{\partial x_2}$$

its vorticity. Since the fluid is incompressible, we have

$$\operatorname{div} \underline{u} = \frac{\partial u_1}{\partial x_1} + \frac{\partial u_2}{\partial x_2} = 0.$$

Hence, it follows from the simple connectivity of Ω that there exists a unique stream function ψ such that

$$\begin{cases} \underline{u} = \underline{\text{curl}}\ \psi = \left(\dfrac{\partial \psi}{\partial x_2}, -\dfrac{\partial \psi}{\partial x_1}\right) \text{ in } \Omega, \\ \psi = 0 \text{ on } \Gamma. \end{cases}$$

This stream function ψ is in fact characterised by

$$\begin{cases} -\Delta\psi = \omega \text{ in } \Omega, \\ \psi = 0 \text{ on } \Gamma. \end{cases} \tag{2.1}$$

If we assume that the vorticity ω is known, the corresponding velocity field $\underline{u}$ is determined by solving the Dirichlet problem (2.1). As it is well known, assuming that ω belongs to the space $L^2(\Omega)$, ψ is also characterised by the variational principle

$$J(\psi) = \min_{\phi \in H_0^1(\Omega)} J(\phi)$$

where

$$J(\phi) = \int_\Omega \{\tfrac{1}{2}|\underline{\text{curl}}\ \phi|^2 - \omega\phi\}\, dx,$$

or equivalently by the variational formulation

$$\begin{cases} \underline{\text{Find}}\ \psi \in H_0^1(\Omega)\ \underline{\text{such that}} \\ \displaystyle\int_\Omega \underline{\text{curl}}\ \psi \cdot \underline{\text{curl}}\ \phi\, dx = \int_\Omega \omega\phi\, dx \qquad \forall \phi \in H_0^1(\Omega) \end{cases} \tag{2.2}$$

Note that, in general, we are more interested in the direct determination of the velocity $\underline{u}$ rather than of the stream function ψ itself. Hence, we want to construct a two-field variational formulation of problem (2.1) which will allow us to define a direct finite element approximation of the velocity field $\underline{u}$. We start from the equations

$$\begin{cases} \underline{u} = \underline{\text{curl}}\ \psi, \\ \text{curl}\ \underline{u} = \omega. \end{cases}$$

By multiplying the first equation by a sufficiently smooth test vector function $\underline{v}$ and using Green's formula over Ω, we obtain

$$\int_\Omega \underline{u}\cdot\underline{v}\ dx = \int_\Omega \underline{\text{curl}}\ \psi\cdot\underline{v}\ dx = \int_\Omega \psi\ \text{curl}\ \underline{v}\ dx.$$

Similarly, by multiplying the second equation by a test function ϕ and integrating over Ω, we find

$$\int_\Omega \text{curl}\ \underline{u}\ \phi\ dx = \int_\Omega \omega\phi\ dx.$$

Now, in order to give a mathematically precise two-field formulation of problem (2.1), we need to introduce the space

$$\underline{H}(\text{curl};\Omega) = \{\underline{v} \in L^2(\Omega)^2;\ \text{curl}\ \underline{v} \in L^2(\Omega)\}.$$

This is a Hilbert space for the norm

$$\|\underline{v}\|_{\underline{H}(\text{curl};\Omega)} = \left\{\|\underline{v}\|^2_{L^2(\Omega)} + \|\text{curl}\ \underline{v}\|^2_{L^2(\Omega)}\right\}^{\frac{1}{2}}.$$

Then, we introduce the following problem: Find a pair $(\underline{u},\psi) \in \underline{H}(\text{curl};\Omega) \times L^2(\Omega)$ such that

$$\begin{cases} \int_\Omega \underline{u}\cdot\underline{v}\ dx - \int_\Omega \text{curl}\ \underline{v}\ \psi\ dx = 0 & \forall \underline{v} \in \underline{H}(\text{curl};\Omega), \\ \int_\Omega \text{curl}\ \underline{u}\ \phi\ dx = \int_\Omega \omega\phi\ dx & \forall\phi \in L^2(\Omega). \end{cases} \tag{2.3}$$

Clearly, if ψ is the solution of problem (2.1), the pair $(\underline{u} = \underline{\text{curl}}\ \psi, \psi)$ is by construction a solution of problem (2.3). It remains to check that it is indeed the only solution; this will be proved in Section 3.

2.2 The Stokes equations

We next consider the more general and interesting problem of determining the flow of an incompressible fluid in a bounded domain Ω of $\mathbb{R}^n$ ($n = 2$ or 3) when the velocity of the fluid is "small" (creeping flow). The flow is then governed by the Stokes equations

$$\begin{cases} -\nu\,\Delta\underline{u} + \text{grad } p = \underline{f} \text{ in } \Omega, \\ \text{div } \underline{u} = 0 \qquad\qquad \text{in } \Omega, \\ \underline{u} = \underline{0} \qquad\qquad\quad \text{on } \Gamma, \end{cases} \tag{2.4}$$

where again $\underline{u} = (u_1, \ldots, u_n)$ is the velocity of the fluid, p is its pressure and $\underline{f}$ represents the body forces.

Define the space

$$\underline{V}(\Omega) = \{\underline{v} \in H_0^1(\Omega)^n;\ \text{div } \underline{v} = 0\}.$$

Then, the velocity field $\underline{u}$ can be characterised by

$$J(\underline{u}) = \min_{\underline{v} \in \underline{V}(\Omega)} J(\underline{v})$$

where

$$J(\underline{v}) = \int_\Omega \{\tfrac{1}{2}|\text{grad } \underline{v}|^2 - \underline{f}\cdot\underline{v}\}\,dx,$$

$$\text{grad } \underline{v} = \left(\frac{\partial v_i}{\partial x_j}\right)_{1 \leqslant i,\, j \leqslant n},$$

or equivalently as the solution of the following problem

$$\begin{cases} \underline{\text{Find } u \in V(\Omega) \text{ such that}} \\ \displaystyle\int_\Omega \text{grad } \underline{u}\cdot\text{grad } \underline{v}\,dx = \int_\Omega \underline{f}\cdot\underline{v}\,dx \qquad \forall\underline{v} \in \underline{V}(\Omega). \end{cases} \tag{2.5}$$

The above classical variational formulation (2.5) of problem (2.4) is not well suited for finite element approximation since it involves the construction of finite element subspaces of $\underline{V}(\Omega)$, i.e., satisfying the constraint div $\underline{v} = 0$. In fact, although such a construction is possible it leads to rather complicated finite elements of limited applicability. Hence, we prefer to derive a two-field variational formulation of problem (2.4) that will lead to the construction of flexible and efficient mixed F.E.M. Starting from the equations (2.4), we multiply the first equation by a sufficiently smooth vector function $\underline{v}$ which vanishes on Γ and we use Green's formula. We obtain

$$\int_\Omega (-\Delta \underline{u} + \text{grad } p)\cdot\underline{v}\, dx = \int_\Omega (\text{grad } \underline{u}\, \text{grad}\cdot\underline{v} - p \text{ div } \underline{v})\, dx$$

$$= \int_\Omega \underline{f}\cdot\underline{v}\, dx.$$

This suggests another variational formulation of the Stokes equations:
Find a pair $(\underline{u},p) \in H_0^1(\Omega)^n \times L^2(\Omega)$ such that

$$\begin{cases} \int_\Omega \text{grad } \underline{u}\cdot\text{grad } \underline{v}\, dx - \int_\Omega p \text{ div } \underline{v}\, dx = \int_\Omega \underline{f}\cdot\underline{v}\, dx & \forall \underline{v} \in H_0^1(\Omega)^n, \\ \int_\Omega q \text{ div } \underline{u}\, dx = 0 & \forall q \in L^2(\Omega). \end{cases} \tag{2.6}$$

In fact, as it is well known, the pressure p is defined up to an additive constant. Thus, for uniqueness, it is more convenient to look for the pressure p in the space

$$L_0^2(\Omega) = \left\{q \in L^2(\Omega);\ \int_\Omega q\, dx = 0\right\}.$$

3. AN ABSTRACT PROBLEM

The two previous examples can be considered as special cases of the following general situation. Let X and M be two Hilbert spaces with norms $\|\cdot\|_X$ and $\|\cdot\|_M$ respectively. We introduce two continuous bilinear forms

$$a(\cdot,\cdot): X \times X \to \mathbb{R},$$

$$b(\cdot,\cdot): X \times M \to \mathbb{R}.$$

Next, we are given two continuous linear forms

$$\ell:\ v \in X \to \langle \ell, v\rangle \in \mathbb{R},$$

$$\chi:\ q \in M \to \langle \chi, q\rangle \in \mathbb{R}$$

and we consider the following problem: Find a pair $(u,p) \in X \times M$ such that

$$\begin{cases} a(u,v) + b(v,p) = \langle \ell, v\rangle & \forall v \in X, \\ b(u,q) = \langle \chi, q\rangle & \forall q \in M. \end{cases} \tag{3.1}$$

In example 2.1, problem (2.3) is indeed of the form (3.1) if we set

$$\begin{cases} X = \underline{H}(\text{curl};\Omega), \ M = L^2(\Omega), \\ a(\underline{u},\underline{v}) = \int_\Omega \underline{u}\cdot\underline{v} \, dx, \\ b(\underline{v},q) = -\int_\Omega \text{curl } \underline{v} \, q \, dx, \\ \ell = 0, \ <\chi,q> = -\int_\Omega \omega \, q \, dx. \end{cases} \tag{3.2}$$

Similarly, in example 2.2, problem (2.5) is also of the form (3.1) if we take

$$\begin{cases} X = H_0^1(\Omega)^n, \quad M = L_0^2(\Omega), \\ a(\underline{u},\underline{v}) = \int_\Omega \text{grad } \underline{u}\cdot\text{grad } \underline{v} \, dx, \\ b(\underline{v},q) = -\int_\Omega \text{div } \underline{v} \, q \, dx, \\ <\ell,\underline{u}> = \int_\Omega \underline{f}\cdot\underline{v} \, dx, \ \chi = 0. \end{cases} \tag{3.3}$$

Let us give sufficient (and almost necessary) conditions for the abstract problem (3.1) to be well posed. We introduce the following subspace V of the space X:

$$V = \{v \in X; \ b(v,q) = 0 \quad \forall q \in M\}. \tag{3.4}$$

Now, we can state a fundamental result due to Brezzi (1974). See also Babuska (1971) for a related result.

<u>Theorem 1</u>. <u>Assume the two following hypotheses:</u>

(i) <u>there exists a constant $\alpha > 0$ such that</u>

$$a(v,v) \geq \alpha \|v\|_X^2 \quad \forall v \in V; \tag{3.5}$$

(ii) there exists a constant $\beta > 0$ such that

$$\sup_{v \in X} \frac{b(v,q)}{\|v\|_X} \geq \beta \|q\|_M \quad \forall q \in M. \tag{3.6}$$

Then, problem (3.1) has a unique solution $(u,p) \in X \times M$.

For the proof which relies on classical results on functional analysis, we refer to the original paper of Brezzi (1974) or to Girault and Raviart (1979a Chapter I, section 4).

Note that, if (u,p) is a solution of problem (3.1), then u belongs to the affine manifold of X

$$V(\chi) = \{v \in X;\ b(v,q) = \langle\chi,q\rangle \quad \forall q \in M\}$$

and satisfies the equation

$$a(u,v) = \langle\ell,v\rangle \quad \forall v \in V. \tag{3.7}$$

Hence, by the Lax-Milgram lemma, the V-ellipticity condition (3.5) ensures the existence and the uniqueness of $u \in V(\chi)$ solution of (3.7) provided that $V(\chi)$ is not empty.

On the other hand, the condition (3.6) can be written in the form

$$\inf_{q \in M} \sup_{v \in X} \frac{b(v,q)}{\|v\|_X \|q\|_M} \geq \beta$$

and is referred to in the literature as the inf-sup condition (or Brezzi's condition or Babuska-Brezzi's condition). It implies that $V(\chi)$ is not empty. Furthermore, $u \in V(\chi)$ being determined by (3.7), the inf-sup condition is necessary and sufficient for the equation

$$b(v,p) = \langle\ell,v\rangle - a(u,v) \quad \forall v \in X$$

to have a unique solution $p \in M$.

Let us next show that Theorem 1 applies in the cases of problems (2.3) and (2.6).

Example 2.1 (continued). Consider first problem (2.3); it follows from (3.2) that in this case

$$V = \{\underline{v} \in L^2(\Omega)^2;\ \text{curl}\ \underline{v} = 0\}.$$

Hence we have

$$a(\underline{v},\underline{v}) = \|\underline{v}\|^2_{L^2(\Omega)} = \|\underline{v}\|^2_{\underline{H}(\mathrm{curl};\Omega)} \qquad \forall \underline{v} \in V$$

so that the V-ellipticity condition holds. On the other hand, the inf-sup condition becomes

$$\sup_{\underline{v} \in \underline{H}(\mathrm{curl};\Omega)} \frac{\int_\Omega \mathrm{curl}\, \underline{v}\, q\, dx}{\|\underline{v}\|_{H(\mathrm{curl};\Omega)}} \geq \beta \|q\|_{L^2(\Omega)} \qquad \forall q \in L^2(\Omega).$$

Thus, let q be in $L^2(\Omega)$ and let $\phi \in H^1_0(\Omega)$ be the solution of the Dirichlet problem

$$\begin{cases} -\Delta\phi = q & \text{in } \Omega, \\ \phi = 0 & \text{on } \Gamma. \end{cases}$$

Then, $\underline{v} = \underline{\mathrm{curl}}\, \phi$ belongs to the space $\underline{H}(\mathrm{curl};\Omega)$ since $\mathrm{curl}\, \underline{\mathrm{curl}}\, \phi = -\Delta\phi = q \in L^2(\Omega)$. Moreover

$$\|\underline{\mathrm{curl}}\, \phi\|_{L^2(\Omega)} \leq c\|q\|_{L^2(\Omega)}$$

and

$$\|\underline{v}\|_{\underline{H}(\mathrm{curl};\Omega)} = (\|\underline{\mathrm{curl}}\, \phi\|^2_{L^2(\Omega)} + \|q\|^2_{L^2(\Omega)})^{\frac{1}{2}} \leq \sqrt{1+c^2}\, \|q\|_{L^2(\Omega)}.$$

With this function $\underline{v}$, we obtain

$$\frac{\int_\Omega \mathrm{curl}\, \underline{v}\, q\, dx}{\|\underline{v}\|_{\underline{H}(\mathrm{curl};\Omega)}} = (1+c^2)^{-\frac{1}{2}} \|q\|_{L^2(\Omega)}$$

and the inf-sup condition holds with $\beta \geq (1+c^2)^{-\frac{1}{2}}$.

Therefore, it follows from Theorem 1 that problem (2.3) has a unique solution $(\underline{u},\psi) \in \underline{H}(\mathrm{curl};\Omega) \times L^2(\Omega)$. Moreover $\psi \in H^1_0(\Omega)$ is the solution of (2.1) and $\underline{u} = \underline{\mathrm{curl}}\, \psi$.

□

Example 2.2 (continued). Consider next problem (2.6). We have

$$V = \underline{V}(\Omega) = \{\underline{v} \in H_0^1(\Omega)^n;\quad \operatorname{div} \underline{v} = 0\}.$$

Since the bilinear form $a(\cdot,\cdot)$ is $H_0^1(\Omega)^n$ - elliptic, i.e.,

$$a(\underline{v},\underline{v}) \geq \alpha \|\underline{v}\|^2_{H^1(\Omega)} \qquad \forall \underline{v} \in H_0^1(\Omega)^n,$$

the V-ellipticity condition holds trivially. The inf-sup condition becomes here

$$\sup_{\underline{v} \in H_0^1(\Omega)^n} \frac{\int_\Omega q \operatorname{div} \underline{v}\, dx}{\|\underline{v}\|_{H^1(\Omega)}} \geq \beta \|q\|_{L^2(\Omega)} \qquad \forall q \in L_0^2(\Omega).$$

Now, one can show - but this is a non trivial result (cf. for instance Temam (1977) or Girault and Raviart (1979a) Chapter I section 3) - that, with any $q \in L_0^2(\Omega)$, it is possible to associate a function $\underline{v} \in H_0^1(\Omega)^n$ with the following properties

$$\begin{cases} \operatorname{div} \underline{v} = q & \text{in } \Omega, \\ \|\underline{v}\|_{H^1(\Omega)} \leq c\|q\|_{L^2(\Omega)}. \end{cases}$$

This, in turn, implies that the inf-sup condition holds with $\beta \geq \frac{1}{c}$.

Again, it follows from Theorem 1 that problem (2.6) has a unique solution $(\underline{u},p) \in H_0^1(\Omega)^n \times L_0^2(\Omega)$, i.e., the Stokes equations (2.4) are well posed! □

4. AN ABSTRACT APPROXIMATION RESULT

We now turn to the finite element approximation of problems based on a two-field variational formulation of the form (3.1). We begin with an abstract general approach.

Let h be a discretization parameter tending to zero. For each h, we are given two finite-dimensional spaces X_h and M_h such that

$$X_h \subset X, \quad M_h \subset M.$$

Now, we consider the following problem which approximates problem (3.1): Find a pair $(u_h,p_h) \in X_h \times M_h$ such that

$$\begin{cases} a(u_h,v_h) + b(v_h,p_h) = \langle \ell, v_h \rangle & \forall v_h \in X_h, \\ b(u_h,q_h) = \langle \chi, q_h \rangle & \forall q_h \in M_h. \end{cases} \tag{4.1}$$

Introduce the subspace of X_h

$$V_h = \{v_h \in X_h;\ b(v_h,q_h) = 0 \quad \forall q_h \in M_h\}. \tag{4.2}$$

Observe that <u>in general</u> $V_h \not\subset V$. Then, we may state our approximation result (cf. Brezzi (1974) or Girault and Raviart (1979a) Chapter II, section 1).

<u>Theorem 2</u>. <u>Assume that the two following conditions hold</u>:

(i) <u>there exists a constant</u> $\alpha_* > 0$ <u>such that</u>

$$a(v_h,v_h) \geq \alpha_* \|v_h\|_X^2 \qquad \forall v_h \in V_h; \tag{4.3}$$

(ii) <u>there exists a constant</u> $\beta_* > 0$ <u>such that</u>

$$\sup_{v_h \in X_h} \frac{b(v_h,q_h)}{\|v_h\|_X} \geq \beta_* \|q_h\|_M \qquad \forall q_h \in M_h. \tag{4.4}$$

<u>Then, problem</u> (4.1) <u>has a unique solution</u> $(u_h,p_h) \in X_h \times M_h$ <u>and we have for some constant</u> $C = C(\alpha_*,\beta_*) > 0$

$$\begin{cases} \|u_h - u\|_X + \|p_h - p\|_M \leq \\ \leq C(\alpha_*,\beta_*) \left\{ \inf_{v_h \in X_h} \|u - v_h\|_X + \inf_{q_h \in M_h} \|p - q_h\|_M \right\}. \end{cases} \tag{4.5}$$

Note that the existence and the uniqueness of the solution (u_h,p_h) of (4.1) follow at once from Theorem 1 (applied with X_h and M_h). On the other hand, when the constants α_*,β_* are <u>independent of h</u>, the bound (4.5) leads to an optimal approximation result. In that case, the conditions (4.3) and (4.4) may be viewed as <u>stability conditions</u>. Together with the <u>consistency conditions</u>

$$\lim_{h \to 0} \inf_{v_h \in X_h} \|u - v_h\|_X = 0, \tag{4.6}$$

$$\lim_{h \to 0} \inf_{q_h \in M_h} \|p - q_h\|_M = 0, \tag{4.7}$$

the stability conditions imply the convergence of the method of approximation, i.e.,

$$\lim_{h \to 0} \left\{ \|u_h - u\|_X + \|p_h - p\|_M \right\} = 0.$$

Since, in general $V_h \not\subset V$, the ellipticity condition (3.5) does not necessarily carry over to V_h. In practice however, the discrete condition (4.3) is usually easily checked with a constant α_* independent of h.

As far as the inf-sup condition is concerned, it is clear that the condition (3.6) does not imply its discrete counterpart (4.4). In fact, (4.4) acts as a compatibility condition between the spaces X_h and M_h which does not hold for arbitrary choices of X_h and M_h. In practice, finding pairs of spaces X_h and M_h satisfying the discrete inf-sup condition (4.4) with a constant β_* independent of h is the crucial problem associated with the use of mixed F.E.M. We shall show in the sequel how to solve this problem in the two examples introduced above.

Remark 1. If the discrete inf-sup condition (4.4) does not hold, one cannot guarantee the existence and uniqueness of p_h and a fortiori the error estimate (4.5). Nevertheless, it may occur that, in some cases, one still obtains good approximation results for u_h. We shall not develop this point which is far beyond the scope of this expository paper; see however Remark 2, section 6. □

Let us briefly discuss the algebraic aspects of problem (4.1). We choose a basis $(w_i)_{1 \leq i \leq I}$ of the space X_h and a basis $(r_j)_{1 \leq j \leq J}$ of the space M_h. If we set

$$u_h = \sum_{i=1}^{I} U_i w_i \quad , \quad p_h = \sum_{j=1}^{J} P_j r_j,$$

the equations (4.1) are equivalent to the $(I+J) \times (I+J)$ linear system

$$\begin{pmatrix} A & B^T \\ B & 0 \end{pmatrix} \begin{pmatrix} U \\ P \end{pmatrix} = \begin{pmatrix} C \\ D \end{pmatrix}$$

where the $I \times I$ matrix A and the $J \times I$ matrix B are defined by

$$A = (a(w_j, w_i))_{1 \leq i,j \leq I},$$

$$B = (b(w_i, r_j))_{\substack{1 \leq i \leq I, \\ 1 \leq j \leq J}}$$

while the I-vectors U,C and the J-vectors P,D are defined by

$$U = (U_i)_{1 \leq i \leq I}, \quad C = (<\ell, w_i>)_{1 \leq i \leq I},$$

$$P = (P_j)_{1 \leq j \leq J}, \quad D = (<\chi, r_j>)_{1 \leq j \leq J}.$$

In many cases, the matrix A is symmetric. Note that the conditions (4.3) and (4.4) (with the constants α_* and β_* possibly depending on h) may be written respectively in the equivalent forms

$$U^T A\, U > 0 \qquad \forall U \in \text{Ker}(B), \tag{4.9}$$

$$\text{Ker}(B^T) = \{0\}. \tag{4.10}$$

It remains to design efficient numerical methods of solution of linear systems of the form (4.8) which satisfy the conditions (4.9) and (4.10). In that direction, we refer to Fortin and Glowinski (1983) where various methods of solution of (4.8) are analysed and compared.

5. MIXED FINITE ELEMENT APPROXIMATION OF THE DIRICHLET PROBLEM FOR THE LAPLACE OPERATOR

Consider again problem (2.1) and define a mixed F.E.M. of approximation based on the two-field variational formulation (2.3). We have to construct finite-dimensional subspaces X_h of $X = \underline{H}(\text{curl};\Omega)$ and M_h of $M = L^2(\Omega)$ such that the discrete inf-sup condition holds.

Assume for simplicity that the two-dimensional domain Ω is polygonal. Introduce a triangulation τ_h of $\bar{\Omega}$ with triangles T whose diameters are $\leq h$ and which satisfy the angle condition (i.e., $\theta \geq \theta_0$ for any angle θ of the triangulation τ_h). Using the terminology of Ciarlet (1978), we have a regular triangulation of $\bar{\Omega}$.

We begin by constructing the subspace X_h of $\underline{H}(\text{curl};\Omega)$. Given a triangle $T \in \tau_h$, we denote by $\underline{\nu} = \underline{\nu}_T$ the unit outward normal and by $\underline{\tau} = \underline{\tau}_T$ the unit tangent along the boundary ∂T of T. Now, given a vector function $\underline{v}$ whose restriction to any triangular element $T \in \tau_h$ is smooth, we observe that $\underline{v}$ belongs to the space $\underline{H}(\text{curl};\Omega)$ if and only if its tangential component $\underline{v}\cdot\underline{\tau}$ is continuous at the interelement boundaries (the normal component $\underline{v}\cdot\underline{\nu}$ need not satisfy any continuity requirement).

Let us then describe the simplest possible choice for the space X_h. We require that a function $\underline{v}_h \in X_h$ satisfies the two following conditions in each triangle $T \in \tau_h$:

(i) $\underline{v}_h\cdot\underline{\tau}$ is constant on each side of ∂T;

(ii) curl $\underline{v}_h$ is constant in T.

In order to make things more precise, we first consider the case of the unit triangle $\hat{T}$ in the (ξ_1,ξ_2)-plane whose vertices are (0,0), (1,0), (0,1). Then, we denote by $\hat{\underline{X}}$ the space of vector functions $\hat{v}$ of the form

$$\begin{cases} \hat{v}_1 = a\xi_2 + b, \\ \hat{v}_2 = -a\xi_1 + c. \end{cases} \tag{5.1}$$

Note that a function $\hat{\underline{v}} \in \hat{\underline{X}}$ satisfies the above requirements in $\hat{T}$ and its degrees of freedom may be chosen as the values of its tangential component $\underline{v}\cdot\underline{\tau}$ at the mid-points of the sides of $\partial\hat{T}$.

Next, let T be any triangle of τ_h and $F = F_T$ an affine invertible mapping of $\mathbb{R}^2$ which maps $\hat{T}$ onto T. We denote by $J = J_T$ the Jacobian determinant of the mapping F_T. Then, we introduce the following transformation $G = G_T : \hat{\underline{v}} \to \underline{v} = G\,\hat{\underline{v}}$ which "preserves" the curl and the tangential components of vector fields

$$\begin{cases} v_1 \circ F = \dfrac{1}{J}\left(\hat{v}_1 \dfrac{\partial F_2}{\partial \xi_2} - \hat{v}_2 \dfrac{\partial F_2}{\partial \xi_1}\right) \\[2ex] v_2 \circ F = \dfrac{1}{J}\left(-\hat{v}_1 \dfrac{\partial F_1}{\partial \xi_2} + \hat{v}_2 \dfrac{\partial F_1}{\partial \xi_1}\right) \end{cases} \tag{5.2}$$

and we set

$$\underline{X}_T = \{\underline{v} : T \to \mathbb{R}^2;\ \underline{v} = G\,\hat{\underline{v}}\ ,\ \hat{\underline{v}} \in \hat{\underline{X}}\}. \tag{5.3}$$

Again, a function $\underline{v} \in \underline{X}_T$ is uniquely determined by the values of its tangential component $\underline{v}\cdot\underline{\tau}$ at the midpoints of the sides of ∂T.

Finally, we take

$$X_h = \left\{ \underline{v}_h \in \underline{H}(\mathrm{curl};\Omega);\ \underline{v}_h|_T \in \underline{X}_T \quad \forall T \in \tau_h \right\}. \tag{5.4}$$

Equivalently, a function $\underline{v}$ belongs to X_h if and only if the two following conditions hold:

(i) $\underline{v}|_T \in \underline{X}_T$ for all $T \in \tau_h$;

(ii) $\underline{v}|_{T_1} \cdot \underline{\nu}_{T_1} + \underline{v}|_{T_2} \cdot \underline{\nu}_{T_2} = 0$ on $\partial T_1 \cap \partial T_2$ for all pairs (T_1, T_2) of adjacent triangular elements of τ_h.

As a consequence, the degrees of freedom of a function $\underline{v} \in X_h$ may be chosen as the values of the tangential component of $\underline{v}$ at the mid-points of the edges of the triangulation τ_h.

Let us next define the subspace M_h of $L^2(\Omega)$. We set:

$$M_h = \left\{ q_h \in L^2(\Omega);\ q_h|_T \in P_0(T) \quad \forall T \in \tau_h \right\} \tag{5.5}$$

where, for any integer $m \geq 0$, $P_m(T)$ denotes the space of the restrictions to the triangle T of all polynomials of degree $\leq m$.

Note that curl $\underline{v}_h \in M_h$ if $\underline{v}_h \in X_h$. Hence, we have

$$V_h = \left\{ \underline{v}_h \in X_h\ ;\ \mathrm{curl}\ \underline{v}_h = 0 \right\} \subset V$$

so that the discrete V_h-ellipticity property (4.3) holds trivially in that case. On the other hand, the discrete inf-sup condition (4.4) is technically difficult to check. In fact, this is a consequence of the following property whose proof can be found in Raviart and Thomas (1977): With any $q_h \in M_h$, one can associate a function $\underline{v}_h \in X_h$ such that

$$\begin{cases} \mathrm{curl}\ \underline{v}_h = q_h, \\ \|\underline{v}_h\|_{\underline{H}(\mathrm{curl};\Omega)} \leq C \|q_h\|_{L^2(\Omega)}, \end{cases} \tag{5.6}$$

where the constant $C > 0$ is independent of h.

Now, by Theorem 2, there exists a unique pair $(u_h, \psi_h) \in X_h \times M_h$ which solves

$$\begin{cases} \int_\Omega \underline{u}_h \cdot \underline{v}_h \, dx - \int_\Omega \text{curl } \underline{v}_h \, \psi_h \, dx = 0 & \forall \underline{v}_h \in X_h, \\ \int_\Omega \text{curl } \underline{u}_h \, \phi_h \, dx = \int_\Omega \omega \, \phi_h \, dx & \forall \phi_h \in M_h. \end{cases} \tag{5.7}$$

Moreover, it follows from (4.5) and standard results in finite element approximation theory that

$$\|\underline{u}_h - \underline{u}\|_{\underline{H}(\text{curl};\Omega)} + \|\psi_h - \psi\|_{L^2(\Omega)} \leq C(\psi)h \tag{5.8}$$

provided that the solution ψ of problem (2.1) is sufficiently smooth.

For the proofs and generalizations to higher order elements, we refer to Raviart and Thomas (1977, 1979) and Falk and Osborn (1980).

6. FINITE ELEMENT APPROXIMATION OF THE STOKES EQUATIONS. A FIRST APPROACH

We now come back to the Stokes equations (2.4). We start from the two-field variational formulation (2.6). In order to define a mixed finite element approximation of the Stokes equations, we have again to construct finite-dimensional subspaces X_h of $X = H_0^1(\Omega)^n$ and M_h of $M = L_0^2(\Omega)$ such that the inf-sup condition

$$\inf_{\underline{v}_h \in X_h} \frac{\int_\Omega q_h \, \text{div } \underline{v}_h \, dx}{\|\underline{v}_h\|_{H^1(\Omega)^n}} \geq \beta_* \|q_h\|_{L^2(\Omega)} \qquad \forall q_h \in M_h \tag{6.1}$$

holds with a constant $\beta_* > 0$ independent of h.

For simplicity, we shall restrict ourselves to the case of the dimension $n = 2$ and, as in Section 5, we shall assume that Ω is a polygonal bounded domain. We introduce a <u>regular</u> triangulation τ_h of $\bar{\Omega}$ made with triangles T whose diameters are $\leq h$. Now, the simplest possible choice of the spaces X_h and M_h consists in setting

$$\begin{cases} X_h = \left\{ \underline{v}_h \in C^0(\bar{\Omega})^2; \; \underline{v}_h|_T \in P_1(T)^2 \; \forall T \in \tau_h, \; \underline{v}_h|_\Gamma = \underline{0} \right\}, \\ M_h = \left\{ q_h \in L_0^2(\Omega); \; q_h|_T \in P_0(T) \quad \forall T \in \tau_h \right\}. \end{cases} \tag{6.2}$$

Observe that we do not require any interelement continuity for the approximate pressures!

Unfortunately, one can easily check that the space

$$V_h = \left\{ \underline{v}_h \in X_h ;\ \int_\Omega q_h \operatorname{div} \underline{v}_h \, dx = 0 \quad \forall q_h \in M_h \right\}$$

is in general reduced to $\{\underline{0}\}$ which is indeed a poor approximation of the space

$$V = \left\{ \underline{v} \in H_0^1(\Omega)^2 ;\ \operatorname{div} \underline{v} = 0 \right\} !$$

Hence, the discrete inf-sup condition clearly cannot hold in that case. The remedy consists in enriching the space X_h of approximate velocities. This can be done in two directions:

(i) with discontinuous approximate pressures (and with penalty techniques);

(ii) with continuous approximate pressures.

6.1 Finite element approximation using discontinuous pressures

Let us first introduce a useful simple condition which implies the discrete inf-sup condition (6.1).

Lemma 1. Assume that there exists a linear continuous operator $\pi_h : H_0^1(\Omega)^n \to X_h$ and a constant $C > 0$ independent of h such that for all $\underline{v} \in H_0^1(\Omega)^n$

$$\int_\Omega q_h \operatorname{div}(\underline{v} - \pi_h \underline{v}) dx = 0 \qquad \forall q_h \in M_h, \tag{6.3}$$

$$\| \pi_h \underline{v} \|_{H^1(\Omega)^n} \leq C \, \| \underline{v} \|_{H^1(\Omega)^n}. \tag{6.4}$$

Then, the discrete inf-sup condition (6.1) holds with a constant $p_* > 0$ independent of h.

Proof. Let q_h be in M_h; by the results of Section 3, we may associate with q_h a function $\underline{v} \in H_0^1(\Omega)^n$ such that

$$\begin{cases} \operatorname{div} \underline{v} = q_h & \text{in } \Omega, \\ \| \underline{v} \|_{H^1(\Omega)} \leq C' \, \| q_h \|_{L^2(\Omega)} . \end{cases}$$

Now, choosing $\underline{v}_h = \pi_h \underline{v}$, we have

$$\int_\Omega q_h \operatorname{div} \underline{v}_h \, dx = \int_\Omega q_h \operatorname{div} \underline{v} \, dx = \|q_h\|^2_{L^2(\Omega)}$$

and the condition (6.1) holds with $\beta_* = \frac{1}{CC'}$. □

As a consequence of Lemma 1, let us show that the choice

$$\left\{\begin{array}{l} X_h = \left\{\underline{v}_h \in C^0(\bar{\Omega})^2;\ \underline{v}_h|_T \in P_2(T)^2\ \forall T \in \tau_h,\ \underline{v}_h|_\Gamma = \underline{0}\right\}, \\ \\ M_h = \left\{q_h \in L^2_0(\Omega);\ q_h|_T \in P_0(T)\ \forall T \in \tau_h\right\} \end{array}\right. \tag{6.5}$$

is now adequate. First, since the bilinear form $a(\cdot,\cdot)$ satisfies

$$a(\underline{v},\underline{v}) = \int_\Omega \operatorname{grad} \underline{v} \cdot \operatorname{grad} \underline{v} \, dx \geq \alpha \|\underline{v}\|^2_{H^1(\Omega)} \qquad \forall \underline{v} \in H^1_0(\Omega)^n,$$

the discrete ellipticity condition holds in any case with $\alpha_* = \alpha$.

On the other hand, in order to apply Lemma 1, we construct the linear operator π_h as follows. Given $\underline{v} \in H^1_0(\Omega)^2$, we denote by $P_h\underline{v}$ its $H^1_0(\Omega)^2$-projection on X_h defined by

$$a(\underline{v} - P_h\underline{v}, \underline{w}_h) = 0 \qquad \forall \underline{w}_h \in X_h.$$

Then, given a triangle $T \in \tau_h$ with vertices a_i, $1 \leq i \leq 3$, we uniquely define $\pi_h\underline{v}|_T \in P_2(T)^2$ by

$$\left\{\begin{array}{l} \pi_h\underline{v}|_T\,(a_i) = P_h\underline{v}(a_i),\ 1 \leq i \leq 3, \\ \\ \int_{[a_i,a_j]} (\pi_h\underline{v}|_T - \underline{v})\,ds = 0,\quad 1 \leq i \leq j \leq 3, \end{array}\right. \tag{6.6}$$

where $[a_i,a_j]$ denotes the side of ∂T with endpoints a_i and a_j. Such a function $\pi_h\underline{v}$ belongs indeed to X_h. Moreover, using Green's formula, we have

$$\int_T \operatorname{div}\,(\pi_h \underline{v} - \underline{v})\,dx = \int_{\partial T} (\pi_h \underline{v} - \underline{v})\cdot\underline{\nu}\,ds = 0$$

so that (6.3) holds. The estimate (6.4) is more difficult to obtain and is proved in Girault and Raviart (1979a) Chapter II, Section 2.2 when Ω is convex. In fact, by replacing in (6.6) $P_k\underline{v}(a_i)$ by a local projection of $\underline{v}$ at the point a_i as in Clement (1975), one can prove the bound (6.4) in any case.

Thus, in the case (6.5), there exists by Lemma 1 and Theorem 2 a unique pair $(\underline{u}_h, p_h) \in X_h \times M_h$ which solves

$$\begin{cases} \displaystyle\int_\Omega \operatorname{grad} \underline{u}_h \cdot \operatorname{grad} \underline{v}_h\,dx - \int_\Omega p_h \operatorname{div} \underline{v}_h\,dx = \\ \qquad\qquad\qquad\qquad = \displaystyle\int_\Omega \underline{f}\cdot\underline{v}_h\,dx \qquad \forall \underline{v}_h \in X_h, \\ \displaystyle\int_\Omega q_h \operatorname{div} \underline{u}_h\,dx = 0 \qquad \forall q_h \in M_h. \end{cases} \tag{6.7}$$

Moreover, if the functions $\underline{u}$ and p are sufficiently smooth ($\underline{u} \in H^2(\Omega)^2$, $p \in H^1(\Omega)$), we have the error bound

$$\|\underline{u}_h - \underline{u}\|_{H^1(\Omega)} + \|p_h - p\|_{L^2(\Omega)} \leqslant C(\underline{u},p)h. \tag{6.8}$$

Hence, we obtain only an $O(h)$ estimate for the error. Clearly, this comes from the coarseness of the space M_h which yields a poor approximation of the incompressibility condition $\operatorname{div} \underline{u} = 0$. Let us therefore introduce a more satisfactory choice of the pair of spaces (X_h, M_h). Given a triangle T, we denote by $\lambda_i = \lambda_i(x)$ the barycentric coordinates of a point $x \in \mathbb{R}^2$ with respect to the vertices of T and we set

$$P(T) = P_2(T) \oplus \lambda_1\lambda_2\lambda_3 = \operatorname{span}\left\{\lambda_1^2, \lambda_2^2, \lambda_3^2, \lambda_1\lambda_2, \lambda_2\lambda_3, \lambda_3\lambda_1, \lambda_1\lambda_2\lambda_3\right\}.$$

Next, we choose

$$\begin{cases} X_h = \left\{\underline{v}_h \in C^0(\bar{\Omega})^2;\ \underline{v}_h|_T \in P(T)^2 \quad \forall T \in \tau_h;\ \underline{v}_h|_\Gamma = \underline{0}\right\}, \\ M_h = \left\{q_h \in L_0^2(\Omega);\ q_h|_T \in P_1(T) \quad \forall T \in \tau_h\right\}. \end{cases} \tag{6.9}$$

Note that the degrees of freedom of a function $\underline{v}_h \in X_h$ can be chosen as the values of $\underline{v}_h$ at the vertices of τ_h and at the midpoints of the edges of τ_h and at the centroids of the triangles of τ_h.

In order to check the discrete inf-sup condition, we construct again a linear operator π_h which satisfies the conditions of Lemma 1. For all triangles $T \in \tau_h$, we uniquely define $\pi_h \underline{v}|_T \in P(T)^2$ by

$$\begin{cases} \pi_h \underline{v}|_T (a_i) = \underline{v}(a_i), \quad 1 \leqslant i \leqslant 3 \\ \int_{[a_i,a_j]} (\pi_h \underline{v}|_T - \underline{v}) ds = 0, \quad 1 \leqslant i \leqslant j \leqslant 3, \\ \int_T x_k \operatorname{div}(\pi_h \underline{v}|_T - \underline{v}) dx = 0, \quad k = 1,2. \end{cases} \tag{6.10}$$

Such a function $\pi_h \underline{v}$ belongs to X_h (i.e., $\pi_h \underline{v}$ is continuous at the inter-element boundaries and vanishes on Γ) and

$$\int_T q \operatorname{div}(\pi_h \underline{v} - \underline{v}) dx = 0 \qquad \forall q \in P_1(T)$$

so that (6.3) holds. The estimate (6.4) is checked as above.

Therefore, in the case (6.9), we get again by Lemma 1 and Theorem 2 the existence and uniqueness of a solution $(\underline{u}_h, p_h)$ of problem (6.7). Moreover, if the functions $\underline{u}$ and p are smooth enough ($\underline{u} \in H^3(\Omega)^2$, $p \in H^2(\Omega)$), we obtain here

$$\|\underline{u}_h - \underline{u}\|_{H^1(\Omega)} + \|p_h - p\|_{L^2(\Omega)} \leqslant C(\underline{u},p)h^2. \tag{6.11}$$

<u>Remark 2</u>. Similar results hold in the case of rectangular elements. Given a rectangle R whose sides are parallel to the axes, we denote by $Q_m(R)$ the space of the restrictions to R of all polynomials of the form

$$\sum_{0 \leqslant \alpha_1, \alpha_2 \leqslant m} C_\alpha x_1^{\alpha_1} x_2^{\alpha_2} \qquad \alpha = (\alpha_1, \alpha_2).$$

If Ω is a rectangle and τ_h is a "quadrangulation" of $\bar{\Omega}$ made with rectangles R whose diameters are $\leqslant h$, the choice

$$\left\{\begin{array}{l} X_h = \left\{ \underline{v}_h \in C^0(\bar{\Omega})^2; \underline{v}_h|_R \in Q_2(R)^2 \quad \forall R \in \tau_h, \ \underline{v}_h|_\Gamma = \underline{0} \right\} \\ \\ M_h = \left\{ q_h \in L_0^2(\Omega); \ q_h|_R \in P_1(R) \quad \forall R \in \tau_h \right\} \end{array}\right. \tag{6.12}$$

(i.e., biquadratic velocities and linear pressures) can be analysed exactly as in the case (6.9) and leads to similar conclusions.

On the other hand, if we take

$$\left\{\begin{array}{l} X_h = \left\{ \underline{v}_h \in C^0(\bar{\Omega})^2; \ \underline{v}_h|_R \in Q_1(R)^2 \quad \forall R \in \tau_h, \ \underline{v}_h|_\Gamma = \underline{0} \right\} \\ \\ M_h = \left\{ q_h \in L_0^2(\Omega); \ q_h|_R \in P_0(R) \quad \forall R \in \tau_h \right\} \end{array}\right. \tag{6.13}$$

(i.e., bilinear velocities and constant pressures: a popular choice!), the inf-sup condition does not hold. However, we still have the existence and the uniqueness of an approximate velocity $\underline{u}_h$ which satisfies the error bound

$$\| \underline{u}_h - \underline{u} \|_{H^1(\Omega)} \leq C(\underline{u}, p) h \tag{6.14}$$

provided that $\underline{u} \in H^2(\Omega)^2$ and $p \in H^1(\Omega)$ (cf. Remark 1). But the approximate pressure is not uniquely defined: there exist spurious modes of checkerboard type which have to be eliminated by filtering in order to obtain a satisfactory numerical answer. For a complete analysis of the case (6.13), we refer to Johnson and Pitkäranta (1982). Let us mention that there exists an increasing literature concerning the discrete inf-sup condition for the Stokes equations and particularly the following problem: What can be expected when this discrete inf-sup condition is violated?

□

For a discussion of two-dimensional and three-dimensional examples of finite elements which satisfy (or "almost" satisfy) the discrete inf-sup condition, we refer for instance to Fortin (1981).

6.2 Finite element approximation using continuous pressures

Another way of improving the mixed F.E.M. defined by (6.5) is to use continuous linear pressures. This leads to the popular method introduced by Hood and Taylor (1973) and associated with the choice

$$\begin{cases} X_h = \left\{ \underline{v}_h \in C^0(\bar{\Omega})^2;\ \underline{v}_h|_T \in P_2(T)^2\ \forall T \in \tau_h,\ \underline{v}_h|_\Gamma = \underline{0} \right\}, \\ \\ M_h = \left\{ q_h \in C^0(\bar{\Omega}) \cap L_0^2(\Omega);\ q_h|_T \in P_1(T)\ \forall T \in \tau_h \right\}. \end{cases} \tag{6.15}$$

In that case, Bercovier and Pironneau (1979) have been able to show that the discrete inf-sup condition (6.1) indeed holds together with the error bound (6.11). Moreover, this F.E.M. is trivially generalized to the three-dimensional Stokes equations. The proof is rather technical and we refer for the details to the paper of Bercovier and Pironneau. For a F.E.M. related to the method of Hood and Taylor, see Glowinski and Pironneau (1979b).

7. THE PENALTY METHOD AND THE REDUCED INTEGRATION TECHNIQUE

Let us go back to the abstract problem (3.1). We shall use a penalty method in order to eliminate the element p and obtain a simpler problem to solve. Denote by $(\cdot,\cdot)$ the inner product of the Hilbert space M. It is convenient here to identify M and its dual space so that $X \in M$ and

$$\langle \chi, q\rangle = (\chi, q).$$

Now, for all $\varepsilon > 0$, we introduce the <u>regularized problem</u>:
<u>Find a pair</u> $(u^\varepsilon, p^\varepsilon) \in X \times M$ <u>such that</u>

$$\begin{cases} a(u^\varepsilon, v) + b(v, p^\varepsilon) = \langle \ell, v\rangle & \forall v \in X, \\ \\ -\varepsilon(p^\varepsilon, q) + b(u^\varepsilon, q) = (\chi, q) & \forall q \in M. \end{cases} \tag{7.1}$$

If we define the linear operator $B \in \mathcal{L}(X;M)$ by

$$(Bv, q) = b(v, q) \qquad \forall v \in X,\ \forall q \in M,$$

the second equation may be equivalently written in the form

$$p^\varepsilon = \frac{1}{\varepsilon}(Bu^\varepsilon - \chi). \tag{7.2}$$

Hence, replacing p^ε by its value (7.2) in the first equation (7.1), we obtain the <u>penalized problem</u>: <u>Find</u> $u^\varepsilon \in X$ <u>such that</u>

$$a(u^\varepsilon, v) + \frac{1}{\varepsilon}(Bu^\varepsilon, Bv) = \langle \ell, v\rangle + \frac{1}{\varepsilon}(\chi, Bv) \qquad \forall v \in X. \tag{7.3}$$

Theorem 3. *Assume that there exists a constant* $\alpha > 0$ *such that*

$$a(v,v) + \|Bv\|_M^2 \geq \alpha\|v\|_X^2 \qquad \forall v \in X \tag{7.4}$$

and that the inf-sup condition (3.6) *holds. Then, problems* (3.1) *and* (7.1) *both have a unique solution. Moreover, we have for all* $\varepsilon \leq \varepsilon_0$ *small enough*

$$\|u^\varepsilon - u\|_X + \|p^\varepsilon - p\|_M \leq C(\ell,\chi)\varepsilon. \tag{7.5}$$

In fact, the hypothesis (7.4) implies the ellipticity condition so that, by Theorem 1, problem (3.1) has indeed a unique solution (u,p). On the other hand, since problem (7.1) is equivalent to (7.2) and (7.3), the sole hypothesis (7.4) ensures that the penalized problem (7.3) has a unique solution $u^\varepsilon \in X$ and therefore that the regularized problem (7.1) has a unique solution $(u^\varepsilon, p^\varepsilon)$. The estimate (7.5) is due to Bercovier (1978); its proof can also be found in Girault and Raviart (1979a) Chapter I, Section 4.3.

Next, applying the above technique to problem (4.1) gives a regularized approximate problem: *Find* $(u_h^\varepsilon, p_h^\varepsilon) \in X_h \times M_h$ *such that*

$$\begin{cases} a(u_h^\varepsilon, v_h) + b(v_h, p_h^\varepsilon) = \langle \ell, v_h\rangle & \forall v_h \in X_h, \\ -\varepsilon(p_h^\varepsilon, q_h) + b(u_h^\varepsilon, q_h) = (\chi, q_h) & \forall q_h \in M_h. \end{cases} \tag{7.6}$$

If we denote by ρ_h the orthogonal projection operator from M onto M_h, the second equation (7.6) becomes

$$p_h^\varepsilon = \frac{1}{\varepsilon}\rho_h\,(Bu_h^\varepsilon - \chi). \tag{7.7}$$

Replacing p_h^ε by its value (7.7) in the first equation (7.8) gives the penalized approximate problem: *Find* $u_h^\varepsilon \in X_h$ *such that*

$$\begin{cases} a(u_h^\varepsilon, v_h) + \frac{1}{\varepsilon}(\rho_h Bu_h^\varepsilon, \rho_h Bv) = \\ \qquad = \langle \ell, v_h\rangle + \frac{1}{\varepsilon}(\rho_h\chi, \rho_h Bv_h) \quad \forall v_h \in X_h. \end{cases} \tag{7.8}$$

If we assume that there exists a constant $\alpha_* > 0$ such that

$$a(v_h,v_h) + \|\rho_h Bv_h\|_M^2 \geq \alpha_* \|v_h\|_X^2 \qquad \forall v_h \in X_h, \tag{7.9}$$

the regularized approximate problem (7.6) has a unique solution $(u_h^\varepsilon, p_h^\varepsilon)$. If in addition the discrete inf-sup condition (4.4) holds, it follows from Theorem 3 that we have for some constant $C = C(\alpha_*, \beta_*, \ell, \chi)$ and for ε small enough

$$\|u_h^\varepsilon - u_h\|_X + \|p_h^\varepsilon - p_h\|_M \leq C(\alpha_*,\beta_*,\ell,\chi)\,\varepsilon. \tag{7.10}$$

It is important to notice at this stage that it is not equivalent to discretize the penalized problem (7.3) or to penalize the discrete problem (4.1). In fact, the first approach leads to consider the following problem: Find $\tilde{u}_h^\varepsilon \in X_h$ solution of

$$\begin{cases} a(\tilde{u}_h^\varepsilon, v_h) + \frac{1}{\varepsilon}(B\tilde{u}_h^\varepsilon, Bv_h) = \\[2ex] \qquad\qquad = \langle \ell, v_h) + \frac{1}{\varepsilon}(\chi, Bv_h) \qquad \forall v_h \in X_h. \end{cases} \tag{7.11}$$

Clearly, (7.11) does not coincide with (7.8). Moreover, the analogue of the bound (7.10) does not hold for $\|\tilde{u}_h^\varepsilon - u_h\|_X$. Therefore, <u>discretizing the penalized problem</u> (7.3) <u>is the wrong way of using the penalty technique</u>.

Let us apply the above considerations to the finite element approximation (6.7) of the Stokes equations (2.4). Given the finite-dimensional subspaces X_h and M_h of $H_0^1(\Omega)^2$ and $L_0^2(\Omega)$ respectively, we assume that

$$M_h = Q_h \cap L_0^2(\Omega),$$

where Q_h is a finite-dimensional subspace of $L^2(\Omega)$ which contains the constant functions. Then, if we denote by ρ_h the orthogonal projector from $L^2(\Omega)$ onto Q_h, we find that ρ_h is also the orthogonal projector from $L_0^2(\Omega)$ onto M_h. Now, the penalized approximate problem (7.8) becomes in this example: <u>Find</u> $\underline{u}_h^\varepsilon \in X_h$ <u>such that</u>

$$\begin{cases} \int_\Omega \operatorname{grad} \underline{u}_h^\varepsilon \cdot \operatorname{grad} \underline{v}_h \, dx + \frac{1}{\varepsilon} \int_\Omega \rho_h \operatorname{div} \underline{u}_h^\varepsilon \cdot \rho_h \operatorname{div} \underline{v}_h dx = \\ \qquad = \int_\Omega \underline{f} \cdot \underline{v}_h dx \qquad \forall \underline{v}_h \in X_h, \end{cases} \tag{7.12}$$

and $p_h^\varepsilon \in M_h$ is given by

$$p_h^\varepsilon = -\frac{1}{\varepsilon} \rho_h \operatorname{div} \underline{u}_h^\varepsilon. \tag{7.13}$$

Moreover, assuming that the discrete inf-sup condition (6.1) holds with a constant $\beta_* > 0$ independent of h, we obtain

$$\|\underline{u}_h^\varepsilon - \underline{u}_h\|_{H^1(\Omega)} + \|p_h^\varepsilon - p_h\|_{L^2(\Omega)} \leq C(\underline{f})\ \varepsilon. \tag{7.14}$$

Let us next show that, in the case of the mixed F.E.M. using discontinuous pressures introduced in Section 6.1, the projection operator ρ_h is a local operator so that $\rho_h \operatorname{div} \underline{v}_h$, $\underline{v}_h \in X_h$, is easy to compute and the penalty method easy to implement. For the sake of simplicity, we shall restrict our discussion to the mixed F.E.M. associated with (6.5) for which

$$Q_h = \left\{ q_h \in L^2(\Omega);\ q_h|_T \in P_0(T) \qquad \forall T \in \tau_h \right\}.$$

Let ρ_T be the orthogonal projector from $L^2(T)$ onto $P_0(T)$, i.e.,

$$\rho_T\, q = \frac{1}{\operatorname{meas}(T)} \int_T q\, dx\ , \quad q \in L^2(T).$$

Then, we have

$$(\rho_h q)|_T = \rho_T q \qquad \forall q \in L^2(\Omega).$$

On the other hand, we notice that the quadrature rule

$$\int_T \phi\, dx \simeq \operatorname{meas}(T)\ \phi(a_T), \tag{7.15}$$

where a_T is the centroid of the triangle T, is exact if $\phi \in P_1(T)$. Since for all $\underline{v}_h \in X_h$

$$\operatorname{div} \underline{v}_h|_T \in P_1(T)$$

and

$$\int_T (\rho_T \operatorname{div} \underline{v}_h - \operatorname{div} \underline{v}_h)\,dx = 0,$$

we obtain

$$\rho_T \operatorname{div} \underline{v}_h = (\operatorname{div} \underline{v}_h)(a_T)$$

and therefore

$$\rho_h \operatorname{div} \underline{v}_h|_T = (\operatorname{div} \underline{v}_h)(a_T) \qquad \forall T \in \tau_h. \tag{7.16}$$

Thus, using (7.16), the equations (7.12) and (7.13) become respectively in the case (6.5)

$$\begin{cases} \int_\Omega \operatorname{grad} \underline{u}_h^\varepsilon \cdot \operatorname{grad} \underline{v}_h\,dx + \frac{1}{\varepsilon} \sum_{T \in \tau_h} \operatorname{meas}(T)(\operatorname{div} \underline{u}_h^\varepsilon \cdot \operatorname{div} \underline{v}_h)(a_T) \\ \qquad = \int_\Omega \underline{f} \cdot \underline{v}_h\,dx \qquad \forall \underline{v}_h \in X_h, \end{cases} \tag{7.17}$$

and

$$p_h^\varepsilon(a_T) = -\frac{1}{\varepsilon}(\operatorname{div} \underline{u}_h^\varepsilon)(a_T) \qquad \forall T \in \tau_h. \tag{7.18}$$

Let us relate this result to the <u>reduced integration technique</u> used by the engineers. In fact, the term

$$\frac{1}{\varepsilon}\int_T \rho_h \operatorname{div} \underline{u}_h^\varepsilon \cdot \rho_h \operatorname{div} \underline{v}_h\,dx$$

may be equivalently obtained by computing the integral

$$\frac{1}{\varepsilon}\int_T \operatorname{div} \underline{u}_h^\varepsilon \cdot \operatorname{div} \underline{v}_h\,dx$$

by means of the quadrature rule (7.15). Since

$$\operatorname{div} \underline{u}_h^\varepsilon \cdot \operatorname{div} \underline{v}_h|_T \in P_2(T)$$

and (7.15) is exact only if $\phi \in P_1(T)$, this is called a reduced integration procedure. This remark can be generalized to some extent: In a number of cases, the good penalty method (7.12) is equivalently obtained by using the wrong form of the penalty method

$$\left\{ \begin{aligned} &\int_\Omega \text{grad}\, \underline{u}_h^\varepsilon \cdot \text{grad}\, \underline{v}_h\, dx + \frac{1}{\varepsilon}\int_\Omega \text{div}\, \underline{u}_h^\varepsilon \cdot \text{div}\, \underline{v}_h\, dx = \\ &\qquad = \int_\Omega \underline{f}\cdot\underline{v}_h\, dx \qquad \forall \underline{v}_h \in X_h \end{aligned} \right.$$

and a suitable reduced integration technique for evaluating the term

$$\frac{1}{\varepsilon}\int_\Omega \text{div}\, \underline{u}_h^\varepsilon \cdot \text{div}\, \underline{v}_h\, dx.$$

In particular, this remark holds for all the examples of Section 6.1 (see Bercovier (1978)).

8. A GENERALIZATION. APPLICATION TO INCOMPRESSIBLE FINITE ELEMENT APPROXIMATIONS OF THE STOKES EQUATIONS

Let us conclude this chapter by discussing briefly a useful generalization of the abstract approach of Sections 3 and 4 which will lead to new methods of approximation of problem (3.1).

Let us first give a weaker formulation of problem (3.1). We introduce two Hilbert spaces $\tilde{X}$ and $\tilde{M}$ such that

$$X \subset \tilde{X}\ ,\ \tilde{M} \subset M$$

with dense and continuous imbeddings. We consider also two continuous bilinear forms

$$\tilde{a}(\cdot,\cdot)\ :\ \tilde{X} \times \tilde{X} \to \mathbb{R}$$

$$\tilde{b}(\cdot,\cdot)\ :\ \tilde{X} \times \tilde{M} \to \mathbb{R}$$

which are extensions of the bilinear forms $a(\cdot,\cdot)$ and $b(\cdot,\cdot)$ in the sense that

$$\tilde{a}(u,v) = a(u,v) \qquad \forall u,v \in X,$$

$$\tilde{b}(v,q) = b(v,q) \qquad \forall v \in X,\ \forall q \in \tilde{M}.$$

In addition, we assume that the linear continuous form ℓ is in fact defined on $\tilde{X}$. Then, we consider the following problem: Find a pair $(u,p) \in \tilde{X} \times \tilde{M}$ such that

$$\begin{cases} \tilde{a}(u,v) + \tilde{b}(v,p) = \langle \ell, v\rangle & \forall v \in \tilde{X}, \\ \tilde{b}(u,q) = \langle \chi, q\rangle & \forall q \in \tilde{M}. \end{cases} \tag{8.1}$$

Let us give sufficient conditions ensuring that both problems (3.1) and (8.1) have the same solution (u,p). We introduce the subspace $\tilde{V}$ of $\tilde{X}$

$$\tilde{V} = \{v \in \tilde{X};\ \tilde{b}(v,q) = 0 \quad \forall q \in \tilde{M}\}. \tag{8.2}$$

Then, we have

Theorem 4. In addition to the above hypotheses, we assume that

(i) $\tilde{V} = V$;

(ii) the bilinear form $\tilde{b}(\cdot,\cdot)$ satisfies a weak inf-sup condition in the sense that there exists a constant $\tilde{\beta} > 0$ such that

$$\sup_{v \in \tilde{X}} \frac{b(v,q)}{\|v\|_{\tilde{X}}} \geq \tilde{\beta}\, \|q\|_M \qquad \forall q \in M. \tag{8.3}$$

Then, if the second argument p of the solution $(u,p) \in X \times M$ of problem (3.1) belongs to the space $\tilde{M}$, the pair (u,p) is the only solution of problem (8.1).

For the proof, we refer to Girault and Raviart (1979a) Chapter III, Section 1.1. Note that the hypothesis (i) together with the V-ellipticity of the bilinear form $a(\cdot,\cdot)$ implies that

$$\tilde{a}(v,v) \geq \tilde{\alpha}\, \|v\|^2_{\tilde{X}} \qquad \forall v \in \tilde{V}, \quad \tilde{\alpha} > 0. \tag{8.4}$$

But the conditions (8.3) and (8.4) alone do not ensure that problem (8.1) is well-posed.

Now, for each h, we are given two finite-dimensional subspaces X_h and M_h such that

$$X_h \subset \tilde{X}, \quad M_h \subset \tilde{M}$$

and, instead of approximating problem (3.1), we approximate problem (8.1): Find a pair $(u_h,p_h) \in X_h \times M_h$ such that

$$\begin{cases} \tilde{a}(u_h,v_h) + \tilde{b}(v_h,p_h) = \langle \ell, v_h\rangle & \forall v_h \in X_h, \\ \tilde{b}(u_h,q_h) = \langle \chi, q_h\rangle & \forall q_h \in M_h. \end{cases} \tag{8.5}$$

Let us point out that an analogue of Theorem 2 still holds in this case, the statement of which being somewhat technical. Thus, we refer for a detailed statement to Brezzi and Raviart (1977) and Girault and Raviart (1979a) Chapter III, Section 1.2.

Next, as an illustration of the above abstract framework, we want to describe briefly a new mixed F.E.M. of approximation of the two-dimensional Stokes equations (2.4) which preserves exactly the incompressibility constraint div $\underline{u} = 0$. Thus, assume $n = 2$ and start from the variational formulation (2.5). By taking into account that div $\underline{u} = 0$, we may write

$$\int_\Omega \text{grad } \underline{u} \cdot \text{grad } \underline{v} \, dx = \int_\Omega \text{curl } \underline{u} \text{ curl } \underline{v} \, dx \qquad \forall \underline{v} \in H_0^1(\Omega)^2.$$

Hence, setting $\omega = \text{curl } \underline{u}$, an equivalent formulation of the Stokes equations consists in finding a triple

$$(\underline{u},\omega,p) \in H_0^1(\Omega)^2 \times L^2(\Omega) \times L_0^2(\Omega)$$

which solves

$$\begin{cases} \int_\Omega (\omega \cdot \text{curl } \underline{v} - p \text{ div } \underline{v}) dx = \int_\Omega \underline{f} \cdot \underline{v} \, dx \qquad \forall \underline{v} \in H_0^1(\Omega)^2, \\ \int_\Omega \{\mu(\text{curl } \underline{u} - \omega) - q \text{ div } \underline{u}\} dx = 0 \qquad \forall (\mu,q) \in L^2(\Omega) \times L_0^2(\Omega). \end{cases} \tag{8.6}$$

Now, assume that the solution $(\underline{u},p)$ of (2.4) satisfies the smoothness properties: $\omega = \text{curl } \underline{u} \in H^1(\Omega)$, $p \in H^1(\Omega)$. Then, by using Green's formula

$$\int_\Omega (\mu \text{ curl } \underline{v} - q \text{ div } \underline{v}) dx =$$

$$= \int_\Omega (\underline{\text{curl}}\ \mu + \underline{\text{grad}}\ q) \cdot \underline{v} \, dx, \quad \forall \underline{v} \in H_0^1(\Omega), \quad \forall (\mu,q) \in H^1(\Omega) \times H^1(\Omega),$$

we find that the triple $(\underline{u},\omega,p)$ satisfies the equations

$$\begin{cases} \int_\Omega (\underline{\text{curl}}\ \omega + \underline{\text{grad}}\ p) \cdot \underline{v} \, dx = \int_\Omega \underline{f} \cdot \underline{v} \, dx \qquad \forall \underline{v} \in L^2(\Omega)^2, \\ \int_\Omega \{\underline{\text{curl}}\ \mu \cdot \underline{u} - \mu\omega + \underline{\text{grad}}\ q \cdot \underline{u}\} dx = 0 \\ \qquad\qquad \forall (\mu,q) \in H^1(\Omega) \times (H^1(\Omega) \cap L_0^2(\Omega)). \end{cases} \tag{8.7}$$

In fact, the formulations (8.6) and (8.7) enter in the abstract framework defined above. We set:

$$X = H_0^1(\Omega)^2 \times L^2(\Omega) \;,\; M = L^2(\Omega) \times L_0^2(\Omega)$$

and

$$a((\underline{u},\omega),(\underline{v},\theta)) = \int_\Omega \omega\theta \, dx \;,\; (\underline{u},\omega),(\underline{v},\theta) \in H_0^1(\Omega)^2 \times L^2(\Omega),$$

$$b((\underline{v},\theta),(\mu,q)) = \int_\Omega \{\mu(\text{curl } \underline{v} - \theta) - q \text{ div } \underline{v}\} \, dx,$$

$$(\underline{v},\theta) \in H_0^1(\Omega)^2 \times L^2(\Omega) \;,\; (\mu,q) \in L^2(\Omega) \times L_0^2(\Omega),$$

$$<\ell,(\underline{v},\theta)> = \int_\Omega \underline{f}\cdot\underline{v} \, dx \;,\; \chi = 0.$$

We have

$$V = \{(\underline{v},\theta) \in H_0^1(\Omega)^2 \times L^2(\Omega) \;;\; \text{curl } \underline{v} = \theta \;,\; \text{div } \underline{v} = 0\}$$

and it is an easy matter to check that all the conditions of Theorem 3.1 are satisfied so that there exists a unique pair

$$((\underline{u},\omega),\; (\lambda,p) \in X \times M$$

which solves

$$\begin{cases} \int_\Omega \{\omega\theta + \lambda(\text{curl } \underline{v} - \theta) - p \text{ div } \underline{v}\} \, dx = \\ \qquad\qquad = \int_\Omega \underline{f}\cdot\underline{v} \, dx \qquad \forall(\underline{v},\theta) \in X, \\ \int_\Omega \{\mu(\text{curl } \underline{u} - \omega) - q \text{ div } \underline{u}\} \, dx = 0 \qquad \forall(\mu,q) \in M. \end{cases} \tag{8.8}$$

Comparing with (8.7), we obtain that $(\underline{u},p)$ is indeed the solution of the Stokes equations and $\omega = \lambda = \text{curl } \underline{u}$.

Next, setting

$$\tilde{X} = L^2(\Omega)^2 \times L^2(\Omega) \;,\; \tilde{M} = H^1(\Omega) \times (H^1(\Omega) \cap L_0^2(\Omega))$$

and

$$\tilde{a}(\cdot,\cdot) = a(\cdot,\cdot),$$

$$\tilde{b}((\underline{v},\theta);(\mu,q)) = \int_\Omega \{\underline{\text{curl}}\; \mu\cdot\underline{v} - \mu\theta + \underline{\text{grad}}\; q\cdot\underline{v}\} \, dx \;,$$

one can again check that the conditions of Theorem 4 hold. Hence, assuming that $\lambda = \text{curl } \underline{u} \in H^1(\Omega)$ and $p \in H^1(\Omega)$ we find that the pair $((\underline{u},\omega),(\lambda,p)) \in \tilde{X} \times \tilde{M}$ is the unique solution of

$$\left\{\begin{aligned} &\int_\Omega \{\omega\theta + \underline{\text{curl}}\ \lambda\cdot\underline{v} - \lambda\theta - p \text{ div } \underline{v}\}dx = \\ &\qquad\qquad = \int_\Omega \underline{f}\cdot\underline{v}\ dx \qquad\qquad \forall(\underline{v},\theta) \in \tilde{X}, \\ &\int_\Omega \{\underline{\text{curl}}\ \mu\cdot\underline{u} - \mu\omega + \underline{\text{grad}}\ q\cdot\underline{u}\}dx = 0 \qquad \forall(\mu,q) \in \tilde{M}. \end{aligned}\right. \tag{8.9}$$

Again, we obtain that the formulations (8.7) and (8.9) are equivalent.

Now, the formulation (8.7) will be used for constructing a mixed finite element approximation of (2.4). First, we assume for simplicity that Ω is a simply connected plane domain. Then, any vector field $\underline{v} \in L^2(\Omega)^2$ has a unique orthogonal decomposition of the form

$$\underline{v} = \underline{\text{curl}}\ \phi + \underline{\text{grad}}\ r\ ,\quad \phi \in H_0^1(\Omega)\ ,\quad r \in H^1(\Omega) \cap L_0^2(\Omega).$$

On the other hand, since div $\underline{u} = 0$, there exists a unique stream-function $\psi \in H_0^1(\Omega)$ such that

$$\underline{u} = \underline{\text{curl}}\ \psi.$$

If we replace $\underline{u}$ in (8.7) by $\underline{\text{curl}}\ \psi$ and if we take $\underline{v} = \underline{\text{curl}}\ \phi$, $\phi \in H_0^1(\Omega)$, we obtain the following problem: <u>Find</u> $(\psi,\omega) \in H_0^1(\Omega) \times H^1(\Omega)$ <u>such that</u>

$$\left\{\begin{aligned} &\int_\Omega \underline{\text{curl}}\ \omega\cdot\underline{\text{curl}}\ \phi\ dx = \int_\Omega \underline{f}\cdot\underline{\text{curl}}\ \phi\ dx \qquad \forall\phi \in H_0^1(\Omega), \\ &\int_\Omega (\underline{\text{curl}}\ \mu\cdot\underline{\text{curl}}\ \psi - \mu\omega)dx = 0 \qquad \forall\mu \in H^1(\Omega). \end{aligned}\right. \tag{8.10}$$

Next, if we take $\underline{v} = \underline{\text{grad}}\ r$, $r \in H^1(\Omega) \cap L_0^2(\Omega)$ in (8.7), we get the problem: <u>Find</u> $p \in H^1(\Omega) \cap L_0^2(\Omega)$ <u>such that</u>

$$\left\{\begin{aligned} &\int_\Omega \underline{\text{grad}}\ p\cdot\underline{\text{grad}}\ r\ dx = \int_\Omega (\underline{f} - \underline{\text{curl}}\ \omega)\cdot\underline{\text{grad}}\ r\ dx \\ &\qquad\qquad\qquad \forall r \in H^1(\Omega) \cap L_0^2(\Omega). \end{aligned}\right. \tag{8.11}$$

Observe that (8.10) is a two-field variational formulation of the Stokes problem using the stream function approach

$$\begin{cases} \Delta^2\psi = \operatorname{curl} \underline{f} \quad \text{in } \Omega, \\ \psi = \dfrac{\partial\psi}{\partial\nu} = 0 \quad \text{on } \Gamma, \end{cases} \tag{8.12}$$

while (8.11) is a variational formulation of the Neumann problem for the pressure

$$\begin{cases} \Delta p = \operatorname{div} \underline{f} \quad \text{in } \Omega, \\ \dfrac{\partial p}{\partial\nu} = \underline{f}\cdot\underline{\nu} - \dfrac{\partial\omega}{\partial\tau} \quad \text{on } \Gamma. \end{cases} \tag{8.13}$$

Let us next introduce our mixed F.E.M. As usual, we assume that the simply connected domain Ω is polygonal and τ_h is a regular triangulation of $\bar{\Omega}$ made with triangles T whose diameters are $\leqslant h$. We consider the finite element spaces

$$\Theta_h = \left\{\theta_h \in C^0(\bar{\Omega});\ \theta_h|_T \in P_k(T) \quad \forall T \in \tau_h\right\},$$

$$\Phi_h = \Theta_h \cap H_0^1(\Omega) = \left\{\phi_h \in \Theta_h;\ \phi_h|_\Gamma = 0\right\}.$$

Then, we define the finite-dimensional subspaces X_h and M_h of X and M respectively by

$$X_h = \left\{(\underline{v}_h,\theta_h);\ \underline{v}_h = \underline{\operatorname{curl}}\ \phi_h + \underline{\operatorname{grad}}\ r_h,\ \ \phi_h \in \Phi_h,\ r_h,\ \theta_h \in \Theta_h\right\}$$

and

$$M_h = \Theta_h \times (\Theta_h \cap L_0^2(\Omega)).$$

Now, as in the continuous case, the corresponding problem (8.5) can be equivalently stated in the following way: <u>Find a pair</u> $(\psi_h,\omega_h) \in \Phi_h \times \Theta_h$ <u>which solves</u>

$$\begin{cases} \displaystyle\int_\Omega \underline{\operatorname{curl}}\ \omega_h\cdot\underline{\operatorname{curl}}\ \phi_h\, dx = \int_\Omega \underline{f}\cdot\underline{\operatorname{curl}}\ \phi_h\, dx \qquad \forall\phi_h \in \Phi_h, \\ \displaystyle\int_\Omega \{\underline{\operatorname{curl}}\ \mu_h\cdot\underline{\operatorname{curl}}\ \psi_h - \mu_h\,\omega_h\}dx = 0 \qquad \forall\mu_h \in \Theta_h, \end{cases} \tag{8.14}$$

and $p_h \in \Theta_h \cap L_0^2(\Omega)$ *solving*

$$\begin{cases} \int_\Omega \underline{\text{grad}}\, p_h \cdot \underline{\text{grad}}\, r_h \, dx = \int_\Omega (\underline{f} - \underline{\text{curl}}\, \omega_h) \cdot \underline{\text{grad}}\, r_h \, dx \\ \qquad\qquad\qquad \forall r_h \in \Theta_h \cap L_0^2(\Omega). \end{cases} \tag{8.15}$$

Setting $\underline{u}_h = \underline{\text{curl}}\, \psi_h$, we have div $\underline{u}_h = 0$, i.e., the approximate velocity $\underline{u}_h$ satisfies exactly the incompressibility constraint. Moreover, the computation of $\underline{u}_h$ is decoupled from that of the approximate pressure p_h.

Concerning the convergence of the mixed finite element approximation (8.14) of the biharmonic problem (8.12), we refer to Ciarlet and Raviart (1974), Scholz (1978), Girault and Raviart (1979a) Chapter III, Section 2, Falk and Osborn (1980), Babuška, Osborn and Pitkäranta (1980). See also the review paper of Glowinski and Pironneau (1979a) for algorithmic considerations.

For the convergence of the method (8.14), (8.15) or related mixed methods of solution of the Stokes problem and their extensions to the Navier-Stokes equations, we refer to Girault and Raviart (1979b, 1982) and Raviart (1982). See also Brezzi, Rappaz and Raviart (1980, 1981a, 1981b) for a general theory of approximation of nonlinear problems and some applications to the finite element solution of the Navier-Stokes equations. Finally, we refer to Nedelec (1980, 1982) for an extension of the mixed F.E.M. (8.14) to three-dimensional problems.

REFERENCES

Babuška, I. (1971) Error bounds in the finite element method, *Numer. Math.*, **16**, 322-333.

Babuška, I., Osborn, J.E. and Pitkäranta, J. (1980) Analysis of mixed methods using mesh dependent norms, *Math. Comp.*, **35**, 1039-1062.

Bercovier, M. (1978) Perturbation of mixed variational problems. Application to mixed finite element methods, *RAIRO Anal. Numer.*, **12**, 211-236.

Bercovier, M. and Pironneau, O. (1979) Error estimates for finite element solution of the Stokes problem in the primitive variables, *Numer. Math.*, **33**, 211-224.

Brezzi, F. (1974) On the existence, uniqueness and approximation of saddle point problems arising from Lagrangian multiples, *RAIRO Anal. Numer.*, **R.2**, 129-151.

Brezzi, F. (1979) Nonstandard finite elements for fourth-order elliptic problems, "Energy Methods in Finite Element Analysis", (R. Glowinski, E.Y. Rodin and O.C. Zienkiewicz, Eds.), Wiley, Chichester, 193-211.

Brezzi, F., Rappaz, J. and Raviart, P.A. (1980) Finite-dimensional approximation of nonlinear problems. Part I: Branches of nonsingular solutions, *Numer. Math.*, **36**, 1-25.

Brezzi, F., Rappaz, J. and Raviart, P.A. (1981a) Finite-dimensional approximation of nonlinear problems. Part II: Limit points, *Numer. Math.*, **37**, 1-28.

Brezzi, F., Rappaz, J. and Raviart, P.A. (1981b) Finite-dimensional approximation of nonlinear problems. Part III: Simple bifurcation points, *Numer. Math.*, **38**, 1-30.

Brezzi, F. and Raviart, P.A. (1977) Mixed finite element methods for fourth-order elliptic equations, "Topics in Numerical Analysis III", (J.J.H. Miller, Ed.), Academic Press, London, 33-56.

Ciarlet, P.G. (1978) The Finite Element Method for Elliptic Problems, North-Holland, Amsterdam.

Ciarlet, P.G. and Raviart, P.A. (1974) A mixed finite element method for the biharmonic equation, "Mathematical Aspects of Finite Elements in Partial Differential Equations", (C. de Boor, Ed.), Academic Press, New York, 125-245.

Clement, P. (1975) Approximation by finite element functions using local regularization, *RAIRO Anal. Numer.*, **R.2**, 77-84.

Falk, R.S. and Osborn, J.E. (1980) Error estimates for mixed methods, *RAIRO Anal. Numer.*, **14**, 249-277.

Fortin, M. (1981) Old and new elements for incompressible flows, *Int. J. Num. Meth. Fluids*, **1**, 347-364.

Fortin, M. and Glowinski, R. (1983) Augmented Lagrangian Methods. Application to the numerical solution of boundary value problems, North-Holland, Amsterdam.

Girault, V. and Raviart, P.A. (1979a) Finite Element Approximation of the Navier-Stokes Equations, Lecture Notes in Mathematics 749, Springer, Berlin.

Girault, V. and Raviart, P.A. (1979b) An analysis of a mixed finite element method for the Navier-Stokes equations, *Numer. Math.*, **33**, 235-271.

Girault, V. and Raviart, P.A. (1982) An analysis of upwind schemes for the Navier-Stokes equations, *SIAM J. Numer. Anal.*, **19**, 312-333.

Glowinski, R. and Pironneau, O. (1979a) Numerical methods for the first biharmonic equation and for the two-dimensional Stokes problem, *SIAM Review*, **21**, 167-212.

Glowinski, R. and Pironneau, O. (1979b) On a mixed finite element approximation of the Stokes problem (I). Convergence of the approximate solution, *Numer. Math.*, **33**, 397-424.

Hood, P. and Taylor, C. (1973) A numerical solution of the Navier-Stokes equations using the finite element technique, *Comp. and Fluids*, **1**, 73-100.

Johnson, C. and Pitkäranta, J. (1982) Analysis of some mixed finite element methods related to reduced integration, *Math. Comp.*, **38**, 375-400.

Nedelec, J.C. (1980) Mixed finite elements in $\mathbb{R}^3$, *Numer. Math.*, **35**, 315-341.

Nedelec, J.C. (1982) Elements finis mixtes incompressibles pour l'equation de Stokes dans $\mathbb{R}^3$, *Numer. Math.*, **39**, 97-112.

Raviart, P.A. (1982) Incompressible finite element methods for the Navier-Stokes equations, *Adv. Water Resources*, **5**, 2-8.

Raviart, P.A. and Thomas, J.M. (1977) A mixed finite element method for second order elliptic-problems, "Mathematical Aspects of Finite Element Methods", (I. Galligani and E. Magenes, Eds.), Lecture Notes in Mathematics 606, Springer, Berlin, 292-315.

Raviart, P.A. and Thomas, J.M. (1979) Dual finite element models for second order elliptic problems, "Energy Methods in Finite Element Analysis", (R. Glowinski, E.Y. Rodin and O.C. Zienkiewicz, Eds.), Wiley, Chichester, 175-191.

Scholz, R. (1978) A mixed method for fourth-order problems using linear elements, *RAIRO Anal. Numer.*, **12**, 85-90.

Temam, R. (1977) Navier-Stokes equations, North-Holland, Amsterdam.

CURVED ELEMENTS

A.R. Mitchell

(Department of Mathematical Sciences, University of Dundee, Dundee)

1. INTRODUCTION

Curved boundaries and interfaces arise in many problems in engineering science. In interior problems a small change in the boundary of the region may not have a large effect on the over-all solution, whereas in exterior problems, where the curved shape completely controls the solution, great care may have to be taken to represent the curved boundary, which is usually given by some sort of mathematical equation or an engineering drawing. In most of what follows we look at curved sides in two dimensions. We have little to say about curved surfaces in three dimensions although the isoparametric method applies in this case. Unfortunately little is known about the substitute surface in the isoparametric case in three dimensional problems.

Unless stated otherwise, we concentrate on the problem consisting of

$$-\left(\frac{\partial^2}{\partial x^2}+\frac{\partial^2}{\partial y^2}\right)u = f \text{ in } R$$

$$u = g \text{ on } \partial R$$

where the region R with boundary ∂R has been covered by a triangular net consisting of straight sided triangles in the interior, and triangles with two straight sides and one curved side round the boundary. We assume that we replace R by R^h and ∂R by ∂R^h, that the boundary conditions are interpolated on the replacement boundary, and that in the subsequent calculations exact integration is used. The approximate finite element solution is denoted by U^h, which is C^0 continuous over R^h.

2. PIECEWISE POLYNOMIAL REPLACEMENT OF THE BOUNDARY: DIRECT METHODS

2.1 ∂R^h is a polygon

This is the simplest replacement procedure where the curved boundary is replaced by a polygon, the vertices of the latter lying on the original boundary. Unfortunately as shown by Berger, Scott and Strang (1972)

$$\|u - U^h\|_{H^1} = \begin{cases} O(h) & r = 1 \\ O(h^{3/2}) & r \geq 2 \end{cases}$$

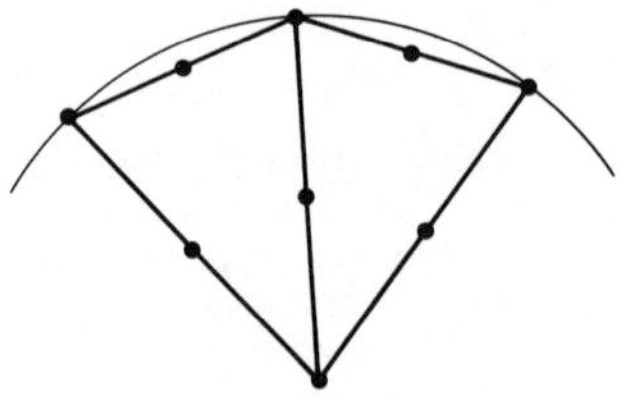

Fig. 1 Polygonal replacement of the boundary

where r is the degree of polynomial, and so higher order polynomials fail to meet the expected rates in the convergence estimates. It should be pointed out, however, that although convergence estimates are useful, they are not the last word since calculations are carried out for a finite value of h which is often quite large when curved boundaries are involved.

2.2 ∂R^h composed of piecewise hyperbolae (McLeod and Mitchell (1972))

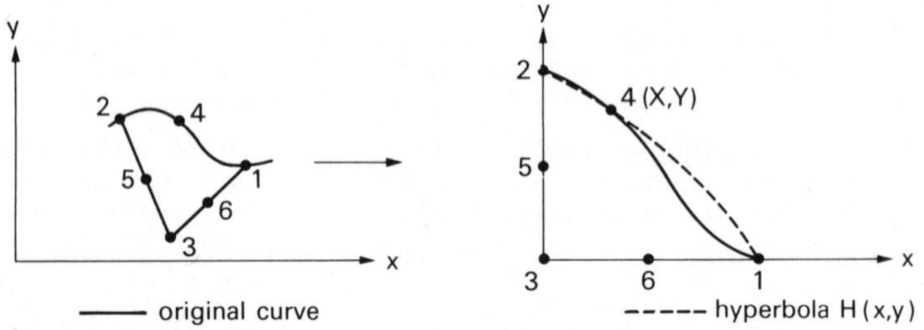

Fig. 2 Replacement of the boundary by piecewise hyperbolae

A triangle with two straight sides and a curved side is transformed by a linear transformation into the triangle shown in Fig. 2, where the points 1 and 2 are unit distance along the x- and y-axes, respectively. The replacement hyperbola through points 1, 2 and 4 is given by

$$H(x,y) \equiv 1 - x - y + \frac{X+Y-1}{XY}\, xy = 0$$

where (X,Y) are the coordinates of the node on the curved side. The Jacobian of the linear transformation is a constant and causes no difficulty. In the transformed plane, (only the quadratic case is illustrated), the approximants in the triangle are

$$U_L(x,y) = x(1 - \frac{y}{Y})U_1 + y(1 - \frac{x}{X})U_2 + HU_3 + \frac{xy}{XY}U_4$$

$$U_Q(x,y) = x(1 - \frac{y}{Y} - 2H)U_1 + y(1 - \frac{x}{X} - 2H)U_2$$

$$+ (1-2x-2y)HU_3 + \frac{xy}{XY}U_4 + 4yHU_5 + 4xHU_6$$

etc., where L, Q, ... represent the linear, quadratic, ... cases, respectively.

All these interpolants lead to C^0 continuity and linear precision, the latter demanding that the basis functions satisfy the relations

$$\sum_{i=1}^{I} B_i = 1, \qquad \sum_{i=1}^{I} B_i x_i = x, \qquad \sum_{i=1}^{I} B_i y_i = y$$

where $I = 4, 6$ in the linear, quadratic, cases, respectively. This time

$$\|u - U^h\|_{H^1} = O(h^r), \quad r = 1,2.$$

2.3 *∂R^h composed of piecewise conics (Rational basis)* (Wachspress (1971, 1973, 1974, 1975))

This time the hyperbola $H(x,y)$ in Fig. 2 is replaced by the conic

$$C(x,y) \equiv 1 - (1+a)x - (1+c)y + ax^2 + bxy + cy^2 = 0$$

with $$b = -\frac{1}{XY}(aX^2 + bY^2 - (1+a)X - (1+c)Y + 1),$$

and the approximants in the triangle become

$$U_L(x,y) = \left(1 - \frac{1 - aX - cY}{1 - ax - cy}\frac{y}{Y}\right)xU_1 + \left(1 - \frac{1 - aX - cY}{1 - ax - cy}\frac{x}{X}\right)yU_2$$

$$+ \frac{C(x,y)}{1 - ax - cy}U_3 + \frac{1 - aX - cY}{1 - ax - cy}\frac{xy}{XY}U_4,$$

etc. All these interpolants lead to C^0 continuity and linear precision and, when $a = c = 0$, this example reduces to that discussed in section 2.2.

3. ISOPARAMETRIC TRANSFORMATIONS

Here the original curved side in the triangle is subjected to a linear transformation (L.T.) as in Fig. 2 and then to a further transformation (isoparametric) (I.T.) to the standard triangle shown in Fig. 3. The complete transformation can be illustrated as

$$(x,y) \overset{\text{L.T.}}{\rightarrow} (x,y) \overset{\text{I.T.}}{\rightarrow} (p,q)$$

and the Jacobian of the over-all transformation as

$$J = J_{\text{L.T.}}\, J_{\text{I.T.}} = \text{Constant} \times J_{\text{I.T.}}$$

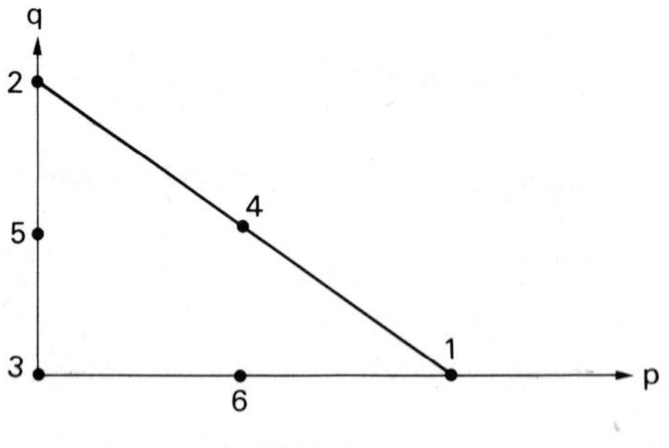

Fig. 3 The standard triangle used for isoparametric transformations

3.1 *Quadratic Case* (illustrated in Figs. 2 and 3)

In the standard triangle in Fig. 3, the quadratic interpolant is

$$U(p,q) = p(2p-1)U_1 + q(2q-1)U_2 + r(2r-1)U_3 + 4pq\,U_4 + 4qr\,U_5 + 4rp\,U_6,$$

where $r = 1-p-q$, leading to the isoparametric transformation T_1 given by

$$\begin{cases} x(p,q) = p(2p-1)x_1+q(2q-1)x_2+r(2r-1)x_3+4pqx_4+4qrx_5+4rpx_6 \\ y(p,q) = \qquad y_1 \qquad y_2 \qquad y_3 \qquad y_4 \qquad y_5 \qquad y_6\ . \end{cases}$$

The construction of the isoparametric transformation ensures that it has linear precision. In the special case when $x_i = p_i$ and $y_i = q_i$ $(i = 1,2,3,5,6)$ and $\alpha = 2(2x_4-1)$, $\beta = 2(2y_4-1)$, T_1 reduces to T_2 given by

$$x = p(1 + \alpha q), \quad y = q(1 + \beta p).$$

Under the transformation T_2, the line $p + q = 1$ becomes a parabola in the (x,y) plane. The parabola can be adjusted by moving the node 4 along the curved side in the original plane. The Jacobian of the transformation T_2 is

$$J = 1 + \alpha q + \beta p$$

and the condition $J > 0$ demands that the point (x_4,y_4) lies in the quadrant $x > \frac{1}{4}$, $y > \frac{1}{4}$ (Mitchell, Phillips and Wachspress (1971)).

3.1.1 Special Cases of the Quadratic Element

(a) Fixed Gradients

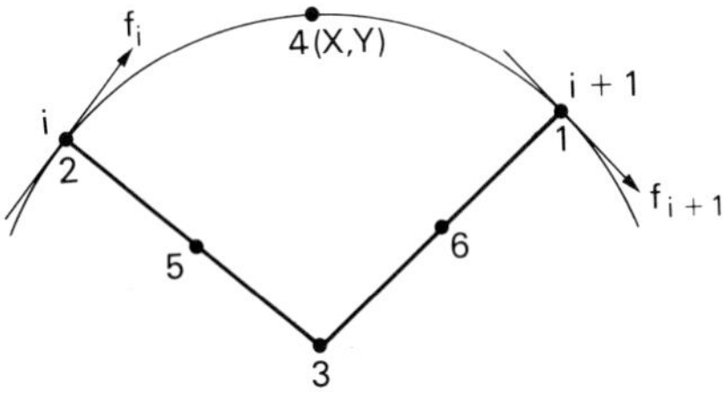

Fig. 4

It is possible to choose the transformation so that the replacement boundary is a C^1 curve. In order to ensure that the isoparametric quadratic curve matches the gradient of the original curve at the end points i and $i + 1$ of the element, we require the coordinates of the node 4 in the original plane to be (see Fig. 4)

$$X = \tfrac{1}{4}\alpha_i + \tfrac{1}{2}(x_i+x_{i+1}), \quad Y = \tfrac{1}{4}\beta_i + \tfrac{1}{2}(y_i+y_{i+1})$$

where

$$\alpha_i = \frac{2(y_{i+1}-y_i) + (x_i-x_{i+1})(f_i+f_{i+1})}{f_i - f_{i+1}}$$

$$\beta_i = \frac{2(x_i-x_{i+1})f_i f_{i+1} - (y_i-y_{i+1})(f_i+f_{i+1})}{f_i - f_{i+1}}$$

with $f_i \neq f_{i+1}$ for a parabolic curve. Further details are available in McLeod and Mitchell (1979).

(b) Four points of parabola coincide with four points of original curve

Normally only three points of the isoparametric parabola coincide with points on the original curve. Special positioning of point 4 off the original curve, however, allows coincidence of four points. This is shown in McLeod and Mitchell (1975). Care must be taken to ensure that $J > 0$.

3.2 Cubic Case (illustrated in Fig. 5)

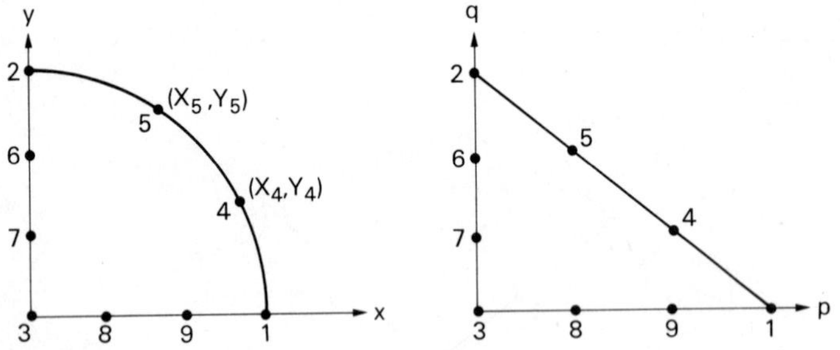

Fig. 5 Cubic isoparametric elements

Here the isoparametric transformation is given by (Mitchell and Wait (1976), p. 97)

$$x(p,q) = p + \frac{9}{2}(6X_{10}-X_4-X_5-1)pq + \frac{27}{2}(X_4-2X_{10})p^2q + \frac{27}{2}(X_5-2X_{10}+\tfrac{1}{3})pq^2$$

$$y(p,q) = q + \frac{9}{2}(6Y_{10}-Y_4-Y_5-1)pq + \frac{27}{2}(Y_4-2Y_{10}+\tfrac{1}{3})p^2q + \frac{27}{2}(Y_5-2Y_{10})pq^2.$$

This time the line p + q = 1 becomes a cubic curve passing through the points 1, 4, 5 and 2 in the (x,y) plane. Unlike the well behaved parabola in the quadratic case, the cubic curve may loop or cusp giving a disastrous replacement of the original curved boundary. If

$$X_5 = X_4 - \frac{1}{3} \qquad Y_5 = Y_4 + \frac{1}{3}$$

$$X_{10} = \frac{1}{2} X_4 \qquad Y_{10} = \frac{1}{2} Y_4 + \frac{1}{6},$$

the cubic curve degenerates into a parabola. Often node 10 is omitted in the cubic case and precision is reduced to second order in the (p,q) plane. Examination of the Jacobian in the cubic case is complicated and no simple rule exists for the placements of nodes 4 and 5 in order to ensure that J > 0. If we follow Woodford, Mitchell and McLeod (1978) with regard to the elimination of node 10, the isoparametric transformation formulae become

$$x = p + pq\,[a + b(p-q)]$$

$$y = q + pq\,[c - d(p-q)]$$

where

$$a = \frac{9}{4}\,(X_4+X_5-1), \qquad b = \frac{27}{4}\,(X_4+X_5-\tfrac{1}{3})$$

$$c = \frac{9}{4}\,(Y_4+Y_5-1), \qquad d = \frac{27}{4}\,(Y_5-Y_4-\tfrac{1}{3})$$

and the Jacobian of the transformation is given by

$$J = 1 + cp + aq - \tfrac{1}{2}dp^2 - \tfrac{1}{2}bq^2 + (b+d)pq + \tfrac{1}{2}(ad+bc)pq(p+q).$$

3.2.1 Analysis of the Jacobian

The object is to find simple rules for the placement of points 4 and 5 on the curved side so that J > 0 for all positions of the point (p,q) in the standard triangle. We give two cases

(i) Symmetric placement of points 4 and 5,

$$\text{i.e.} \qquad x_5 = y_4, \quad y_5 = x_4.$$

This leads to an implied symmetric cubic curve

$$a^3(x-y)^2 = [a + \tfrac{1}{2}b(x+y-1)]^2[a - 2(x + y - 1)]$$

useful for approximating circular arcs (Williams and Morton (1979)). The region for permissible placement of points 4 and 5 is shown in Fig. 6.

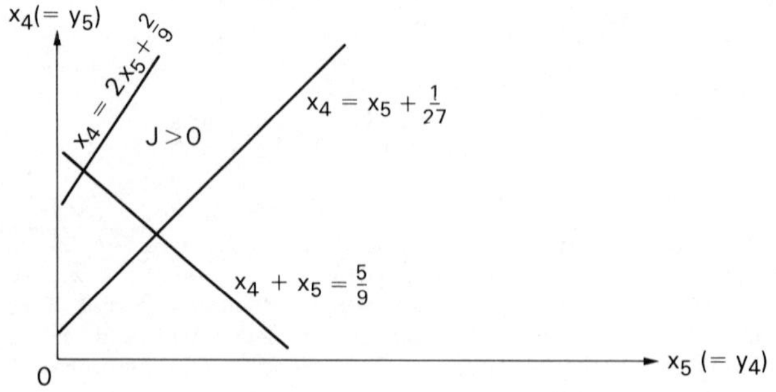

Fig. 6 The placement of nodes 4 and 5 which ensures that J > 0 for the symmetric case ($x_5 = y_4$, $x_4 = y_5$)

(ii) $a = b = 0 \Rightarrow x_4 = \frac{2}{3}, \quad x_5 = \frac{1}{3}.$

The implied curve is now the cubic

$$y = 1 - (1-c-\frac{1}{2}d)x - (c+\frac{3}{d}d)x^2 + dx^3$$

and the permissible region for placement of the points 4 and 5 is given in Woodford, Mitchell and McLeod (1978).

3.2.2 Fixed Gradients

As in the quadratic case, if the slopes of the original curved boundary are f_i and f_{i+1} at the points i and i + 1 on the boundary and we wish the implied cubic curve to match these slopes exactly, then the coordinates of points 4 and 5 must satisfy

$$f_i X_4 - Y_4 - \frac{1}{2} f_i X_5 + \frac{1}{2} Y_5 = \frac{1}{9}[f_i(\frac{11}{2} x_i - x_{i+1}) - (\frac{11}{2} y_i - y_{i+1})]$$

$$\frac{1}{2} f_{i+1} X_4 - \frac{1}{2} Y_4 - f_{i+1} X_5 + Y_5 = \frac{1}{9}[f_{i+1}(x_i - \frac{11}{2} x_{i+1}) - (y_i - \frac{11}{2} y_{i+1})]$$

where i, i+1 are nodes 1, 2, respectively. The above formulae (Mitchell (1979)) allow the replacement boundary to be C^1 continuous which may be desirable in some problems.

4. HIGH ORDER PRECISION

A crucial factor affecting the accuracy of a finite element solution to a problem with curved boundaries is the maximum degree of polynomial spanned by the basis functions of an element adjacent to the boundary. We refer to this as the order or precision of the basis. In the following table we see the respective requirements of linear, quadratic, and cubic precision where $B_i(x,y)$ is the basis function at the node i of the triangle.

Table 1

The conditions necessary for linear(L), quadratic(Q) and cubic(C) precision

L	Q	C
$\Sigma B_i = 1$	$\Sigma B_i x_i^2 = x^2$	$\Sigma B_i x_i^3 = x^3$
$\Sigma B_i x_i = x$	$\Sigma B_i x_i y_i = xy$	$\Sigma B_i x_i^2 y_i = x^2 y$
$\Sigma B_i y_i = y$	$\Sigma B_i y_i^2 = y^2$	$\Sigma B_i x_i y_i^2 = xy^2$
		$\Sigma B_i y_i^3 = y^3$

So far all the methods described have linear precision, and we now look at the possibility of increasing this. In order to succeed, we first require to know how many nodes are required on an algebraic curve of degree m to span polynomials of degree n. The answer is

$$N = \frac{1}{2}[(n+1)(n+2) - (n-m+1)(n-m+2)].$$

Hence for a triangle with two straight sides and a conic arc, the latter being a replacement of the original boundary segment, we require 5 nodes on the conic arc for quadratic precision and 8 nodes for cubic precision. Hence the summation ranges in the precision table are i = 1 to 8 and 1 to 12 in the quadratic and cubic cases, respectively.

4.1 Rational basis with improved precision

Quadratic Case (Fig. 7a) Introduce the conic $C(x,y) = 0$ given in Section 2.3 with nodes 1, 2 and 5 on the original curved boundary. The basis is given by

$$B_7(x,y) = \frac{4y\,C(x,y)}{1 - ax - cy}, \qquad B_8(x,y) = \frac{4x\,C(x,y)}{1 - ax - cy},$$

and the six precision formulae in the Table 1. Nodes 4 and 6 are on the conic but not on the original curved boundary.

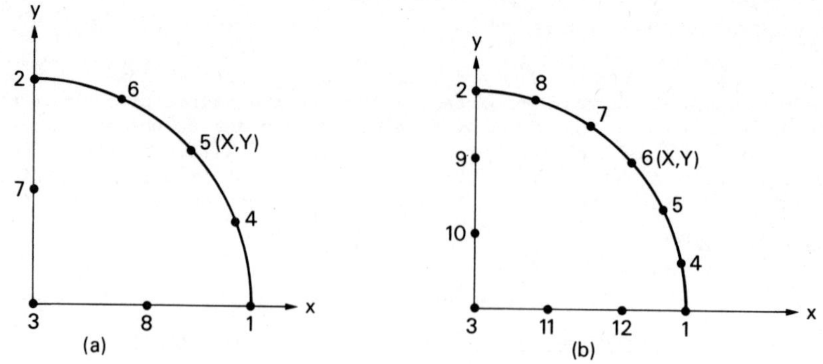

Fig. 7 Rational elements with improved precision

Cubic Case (Fig. 7b) This time the conic has nodes 1, 2 and 6 on the original curve. The basis is obtained from

$$B_{10}(x,y) = \frac{9y(1 - \frac{3}{2}y)C(x,y)}{1 - ax - cy}, \quad B_{11}(x,y) = \frac{9x(1 - \frac{3}{2}x)C(x,y)}{1 - ax - cy},$$

and the ten precision formulae.

4.2 Isoparametric basis with improved precision

Quadratic Case (Fig. 8a) The basis is given by

$$B_4(x,y) = \frac{16}{3}\,pq(6,5)_4, \qquad B_6(x,y) = \frac{16}{3}\,pq(4,5)_6$$

and the six precision formulae, where $(i,j)_k$ is the linear form which vanishes at nodes i and j and takes the value unity at node k, all in the (x,y) plane.

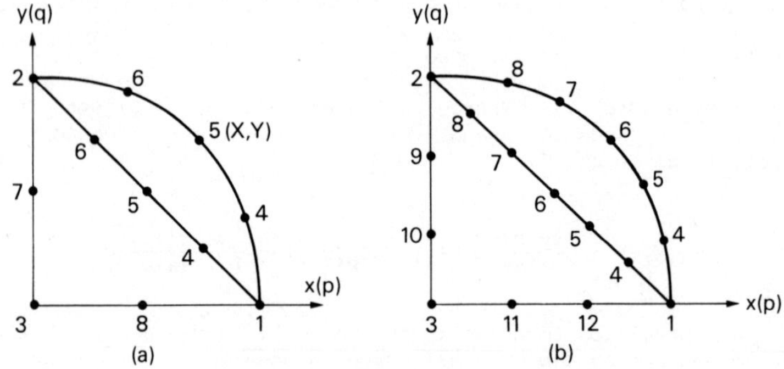

Fig. 8 Isoparametric elements with improved precision: a) quadratic, b) cubic

Cubic Case (Fig. 8b) This is much more complicated and full details are available in McLeod (1978).

5. NUMERICAL EXPERIMENTS

McLeod and Murphy (1979) solved Laplace's equation with Dirichlet boundary conditions in a variety of curved regions. Seven theoretical solutions were considered in five different shapes of curved region. The boundary conditions were read off the theoretical solutions. An L_2 norm of the error was displayed indicating the substantially increased accuracy of the high order methods compared with the standard (linear precision) isoparametric method. The increase in accuracy must of course be balanced against the increase in computational effort.

REFERENCES

Berger, A., Scott, R. and Strang, G. (1972) Approximate boundary conditions in the finite element method. Symposium on Numerical Analysis, Rome: Academic Press.

McLeod, R.J.Y. (1976) High order transformation methods for curved finite elements. *J. Inst. Maths. Applics.*, **21**, 419-428.

McLeod, R.J.Y. and Mitchell, A.R. (1975) The use of parabolic arcs in matching curved boundaries. *J. Inst. Maths. Applics.*, **16**, 259-266.

McLeod, R.J.Y. and Mitchell, A.R. (1972) The construction of basis functions for curved elements in the finite element method. *J. Inst. Maths. Applics.*, **10**, 382-393.

McLeod, R.J.Y. and Mitchell, A.R. (1979) A piecewise parabolic C^1 approximation technique for curved boundaries. *Computers and Maths. with Applics.*, **5**, 277-284.

McLeod, R.J.Y. and Murphy, R. (1979) A numerical comparison of high order transformation and isoparametric transformation methods. *Computers and Maths. with Applics.*, **5**, 241-248.

Mitchell, A.R. (1979) Advantages of cubics for approximating element boundaries. *Computers and Maths. with Applics.*, **5**, 321-328.

Mitchell, A.R., Phillips, G. and Wachspress, E. (1971) Forbidden shapes in the finite element method. *J. Inst. Maths. Applics.*, **8**, 260-269.

Mitchell, A.R. and Wait, R. (1976) The Finite Element Method in Partial Differential Equations. John Wiley and Sons.

Wachspress, E. (1971) A rational basis for function approximation. Lecture Notes in Mathematics. **228**, Springer Verlag.

Wachspress, E. (1973) A rational basis for function approximation II Curved Sides. *J. Inst. Maths. Applics.*, **11**, 83-104.

Wachspress, E. (1974) Conf. Num. Soln. Diff. Equns. Dundee. Lecture Notes in Mathematics, **363**, Springer Verlag.

Wachspress, E. (1975) A Rational Finite Element Basis. Academic Press.

Morton, K.W. and Williams, B.R. (1979) Exterior Flow with an Isoparametric Hermite Cubic Element. *Int. J. Num. Meths. in Engng.*, **14**, 1499-1509.

Woodford, G., Mitchell, A.R. and McLeod, R.Y.J. (1978) An analysis of cubic isoparametric transformation. *Int. J. Num. Meths. in Engng.*, **12**, 1587-1595.

A special issue of Computers and Mathematics with Applications (Pergamon Press) Vol. **5**, No. 4 (1979) is devoted to Curved Finite Elements.

INTRODUCTION TO THE TREATMENT OF SINGULARITIES IN ELLIPTIC BOUNDARY VALUE PROBLEMS USING FINITE ELEMENT TECHNIQUES

J.R. Whiteman

(Institute of Computational Mathematics, Brunel University)

and

K.T. Schleicher

(Fachbereich Mathematik, Technische Hochschule Darmstadt)

1. INTRODUCTION

Many problems in potential theory and linear elasticity occur in regions which contain sharp corners and edges. A considerable body of analysis now exists showing that singularities can occur at such boundary points and lines, with the effect that the regularity of the solution is reduced from that expected for such problems when the regions have smooth boundaries. When singularities occur, whilst the main unknown variable of the problem is important, it is often the case that the strength of the singularity is also of physical significance (e.g. the stress intensity factor in linear elastic fracture).

This paper seeks to outline some of the regularity results for two-dimensional Poisson problems involving singularities and to demonstrate their significance in the analysis of error in finite element methods for this type of problem. In order to do this we consider weak formulations of the problems in a functional analytic setting, indicating how the regularity of the solutions depends on problem characteristics such as the forcing term, the boundary data and the boundary shape; all again functions of physical significance. The finite element method (FEM) is then introduced, together with error bounds. Adaptations of the standard FEM for treating singularities are discussed and analysed. This covers also the approximation of the coefficients of singular terms. Finally a brief comparison is made with boundary element methods (BEM) in the same context, again concentrating on methods for treating the boundary singularities.

2. BOUNDARY VALUE PROBLEMS WITH SINGULARITIES

We consider the mixed boundary value problem

$$\left.\begin{aligned} -\Delta[u(\underline{x})] &= f(\underline{x}), \quad \underline{x} \in \Omega \\ u(\underline{x}) &= 0, \quad \underline{x} \in \Gamma_j,\ j \in \mathcal{D} \neq \emptyset, \\ \frac{\partial u(\underline{x})}{\partial n} &= h_j(\underline{x}), \quad \underline{x} \in \Gamma_j,\ j \in \mathcal{N} \end{aligned}\right\} \qquad (2.1)$$

where $\Omega \subset \mathbb{R}^2$ is a simply connected open polygonal domain with boundary $\partial\Omega \equiv \left(\bigcup_{j\in D} \bar{\Gamma}_j\right) \cup \left(\bigcup_{j\in N} \bar{\Gamma}_j\right) \equiv \partial\Omega_1 \cup \partial\Omega_2$, in which the Γ_j are open straight line segments, $f \in L_2(\Omega)$, and $\partial/\partial n\big|_{\Gamma_j}$ is the derivative in the outward normal direction to Γ_j. The vertices of the boundary are $s_j = \bar{\Gamma}_j \cap \bar{\Gamma}_{j+1}$ at which the internal angles are $\omega_j = \sphericalangle(\Gamma_j, \Gamma_{j+1})$, see Fig. 1. The homogeneous Dirichlet boundary condition in (2.1) has been chosen deliberately. It is assumed that problems with nonhomogeneous Dirichlet conditions will be treated by first being transformed to problems of the form (2.1).

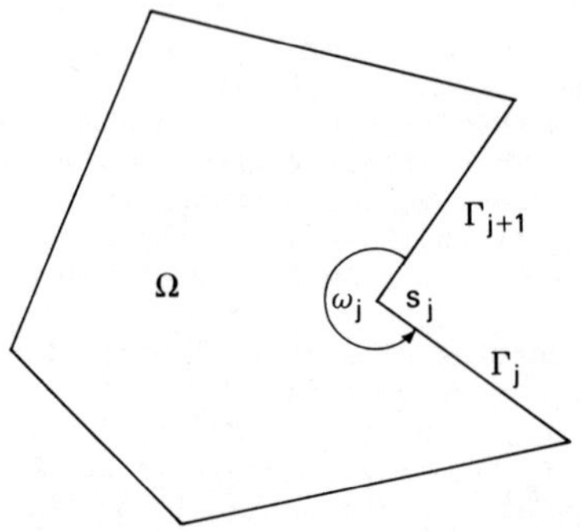

Fig. 1

In the context of error analysis of finite element methods for problems of type (2.1) containing boundary singularities, the regularity of the weak solutions plays an important part. The Sobolev spaces necessary for the weak problems and the error analyses are now defined.

$H^s(\Omega)$, $s = m + \sigma$, m a non-negative integer and $0 \leqq \sigma < 1$, is the space of functions defined in Ω such that

$$\text{for } \sigma = 0, \qquad \|v\|_{H^m(\Omega)} = \left\{ \sum_{|\alpha| \leqq m} \int_\Omega |D^\alpha v(\underline{x})|^2 \, d\underline{x} \right\}^{\frac{1}{2}} < \infty, \tag{2.2}$$

$$\text{for } \sigma \neq 0, \quad \|v\|_{H^s(\Omega)} = \|v\|_{H^m(\Omega)} + \left\{ \sum_{|\alpha|=m} \int_{\Omega\times\Omega} \frac{|D^\alpha v(\xi) - D^\alpha v(\eta)|^2 d\xi d\eta}{|\xi - \eta|^{2+2\sigma}} \right\}^{\frac{1}{2}} < \infty. \tag{2.3}$$

Important specialisations of the above are spaces of functions which satisfy homogeneous boundary conditions a.e. on $\partial\Omega$. Thus we define $H_0^s(\Omega)$ to be the closure of $C_0^\infty(\Omega)$ in $H^s(\Omega)$.

Use is also made later of Sobolev spaces $H_\beta^m(\Omega)$ with *weight* functions $\rho^\beta(\underline{x})$. These weight functions, the complete definition of which is given in Babuska and Rosenzweig (1972), are assumed to have all derivatives in $\bar{\Omega} - \bigcup_{j\in D\cup N} \{s_j\}$. $H_\beta^m(\Omega)$ is defined to be the closure of the class of infinitely differentiable functions on $\bar{\Omega}$ with respect to the norm

$$\|v\|_{H_\beta^m(\Omega)} = \left\{ \sum_{|\alpha|\le m} \int_\Omega |D^\alpha v(\underline{x})|^2 \rho^\beta(\underline{x})\, d\underline{x} \right\}^{\frac{1}{2}} . \qquad (2.4)$$

In the context of problem (2.1) we define the space

$$\tilde{H}^1(\Omega) \equiv \left\{ v : v \in H^1(\Omega) \; ; \; v|_{\partial\Omega_1} = 0 \right\} . \qquad (2.5)$$

Finally spaces on the boundary segments Γ_j are needed. Thus we define

$$H^s(\Gamma_j) \equiv \begin{cases} \left\{ v|_{\Gamma_j} : v \in H^{s+\frac{1}{2}}(\mathbb{R}^2) \right\} & , \; s > 0 \\ L_2(\Gamma_j) & , \; s = 0 . \end{cases} \qquad (2.6)$$

For the case $s > 0$ we suppose that $w \in H^s(\Gamma_j)$ is defined on Γ_j so that $v|_{\Gamma_j} = w$. Then

$$\|w\|_{H^s(\Gamma_j)} = \inf_{v|_{\Gamma_j} = w} \|v\|_{H^{s+\frac{1}{2}}(\mathbb{R}^2)} .$$

The weak form of (2.1) is defined to be:

$$\text{find } u \in \tilde{H}^1(\Omega) : a(u,v) = F(v) \;\; \forall\, v \in \tilde{H}^1(\Omega), \qquad (2.7)$$

where

$$a(u,v) \equiv \int_\Omega \nabla u \, \nabla v \, d\underline{x} \;, \quad u,v \in H^1(\Omega),$$

$$F(v) \equiv \sum_{j\in N} \int_{\Gamma_j} h_j \, v \, ds + \int_\Omega f \, v \, d\underline{x} .$$

It can be shown, see e.g. Gisvard (1982), that the bilinear form a(.,.) and the linear form F(.) fulfil the assumptions of the Lax-Milgram theorem

$$\text{continuity:} \quad |a(u,v)| \leq c_1 \|u\|_{\tilde{H}^1} \|v\|_{\tilde{H}^1} ,$$

$$H^1\text{-ellipticity:} \quad a(u,v) \geq c_2 \|u\|^2_{\tilde{H}^1} , \qquad u, v \in \tilde{H}^1(\Omega),$$

$$|F(v)| \leq c_3 \|v\|_{\tilde{H}^1} ,$$

for constants $c_1, c_2, c_3 > 0$ independent of v. The ellipticity condition can be derived using the Poincaré inequality, since $\mathcal{D} \neq \emptyset$. As a result of the above problem (2.7) has a unique solution.

Kondrat'ev (1967) showed for problems of the type (2.1) that the solution, in cases where Ω has boundary corners, admits a decomposition into a *smooth* part u_μ and a *non-smooth* part, so that

$$u = \sum_{j,\ell} c_{j\ell}\, u_{j\ell} + u_\mu, \tag{2.8}$$

where j is associated with the corner s_j, both the summations are over a finite number of terms, the $c_{j\ell}$ are real numbers, the $u_{j\ell}$ are known *singular* functions and u_μ is a smooth function; (in fact $u_\mu \in H^\mu(\Omega)$).

The work of Grisvard (1976), (1980) considers the weak problem (2.7) in the presence of boundary singularities. In order that the regularity of the weak solution in this case may be determined, spaces *augmented* with a finite number of singular functions are defined as follows. For ℓ a positive integer let

$$\lambda_{j\ell} = \begin{cases} \ell\pi/\omega_j, \text{ if } j,j+1 \in \mathcal{D} \text{ or } j,j+1 \in N, \\ (\ell-\tfrac{1}{2})\pi/\omega_j, \text{ if } j \in \mathcal{D},\ j+1 \in N \text{ or } j \in N,\ j+1 \in \mathcal{D} . \end{cases} \tag{2.9}$$

In terms of coordinates (r_j,θ_j) local to the jth corner (s_j,Γ_{j+1}), as in Fig. 1, we define the functions

$$u_{j\ell}(r_j,\theta_j) = r_j^{\lambda_{j\ell}} \chi_j(r_j)\ \phi_{j\ell}(\theta_j),$$

$$\mathring{u}_{j\ell}(r_j,\theta_j) = r_j^{\lambda_{j\ell}} \chi_j(r_j) [\log r_j \phi_{j\ell}(\phi_j) - \frac{\theta_i}{\lambda_{j\ell}} \frac{d\phi_{j\ell}}{d\theta_j} (\theta_j)] , \tag{2.10}$$

where the χ_j are smooth cut-off functions, $\chi_j = 1$ near s_j, $\chi_j = 0$ away from s_j, and

$$\phi_{j\ell}(\theta_j) = \begin{cases} \sin \lambda_{j\ell}\theta_j, & j,j+1 \in \mathcal{D} \text{ or } j \in \mathcal{D},\ j+1 \in N, \\ \cos \lambda_{j\ell}\theta_j, & j,j+1 \in N, \\ \sin \lambda_{j\ell}(\omega_j-\theta_j), & j \in N,\ j+1 \in \mathcal{D}. \end{cases} \tag{2.11}$$

The augmented space Z_μ is defined for all $\mu \neq \lambda_{j\ell}+1$ so that $v \in Z_\mu \Leftrightarrow$

$$v = v_\mu + \sum_{\substack{j,\ell \\ \lambda_{j\ell} \notin \mathbb{N}}} c_{j\ell}\, u_{j\ell} + \sum_{\substack{j,\ell \\ \lambda_{j\ell} \in \mathbb{N}}} \mathring{c}_{j\ell}\, \mathring{u}_{j\ell}\ , \tag{2.12}$$

where the finite sums are over those values of ℓ for which $0 < \lambda_{j\ell} < \mu - 1$, $v_\mu \in H^\mu(\Omega)$ and the $c_{j\ell}$ and $\mathring{c}_{j\ell}$ are uniquely determined coefficients. The norm for $v \in Z_\mu$ is

$$\|v\|_{Z_\mu} = \|v_\mu\|_{H^\mu(\Omega)} + \sum_{\substack{j,\ell \\ \lambda_{j\ell} \notin \mathbb{N}}} |c_{j\ell}| + \sum_{\substack{j,\ell \\ \lambda_{j\ell} \in \mathbb{N}}} |\mathring{c}_{j\ell}|\ , \tag{2.13}$$

where again $0 < \lambda_{j\ell} < \mu - 1$, and $\mathbb{N}$ is the set of positive integers.

The following theorem due to Grisvard (1982) determines the regularity of the solution u of (2.7).

Theorem 2.1

Let u be the solution of (2.7), where Ω is a general polygonal domain, with $f \in H^\mu(\Omega)$ and $h_j \in H^{\mu+\frac{1}{2}}(\Gamma_j)$, $j \in N$, then $u \in Z_{\mu+2}(\Omega)$, where $Z_{\mu+2}$ was defined in (2.12) with norm (2.13), and

$$\|u\|_{Z_{\mu+2}(\Omega)} \leq C \left\{ \|f\|_{H^\mu(\Omega)} + \sum_{j \in N} \|h_j\|_{H^{\mu+\frac{1}{2}}(\Gamma_j)} \right\}. \tag{2.14}$$

For completeness it must be pointed out that, for a limited number of special corner angles ω_j, the h_j must satisfy the following additional conditions:

(i) $h_j \underset{s_j}{\sim} 0$ $\quad j \in \mathcal{N},\ j+1 \in \mathcal{D},\ w_j = \pi/2,\ 3\pi/2$

(ii) $h_{j+1} \underset{s_j}{\sim} 0$ $\quad j \in \mathcal{D},\ j+1 \in \mathcal{N},\ w_j = \pi/2,\ 3\pi/2$

(iii) $h_j \underset{s_j}{\sim} h_{j+1}$ $\quad j,\ j+1 \in \mathcal{N}\ ,\ w_j = \pi$

(iv) $h_j \underset{s_j}{\sim} -h_{j+1}$ $\quad j,\ j+1 \in \mathcal{N}\ ,\ w_j = 2\pi$

where $g_j \underset{s_j}{\sim} g_{j+1}$ is defined as

$$\int_0^{\delta_j} |g_{j+1}(x_j(\sigma)) - g_j(x_j(-\sigma))|^2 \frac{d\sigma}{\sigma} < \infty$$

for some $\delta_j > 0$, where $x_j(\sigma) \in \partial\Omega$ with σ the distance along $\partial\Omega$ from the corner s_i, $\sigma < 0$ on Γ_j, $\sigma > 0$ on Γ_{j+1}.

As an example of Theorem 2.1 we consider first the case of Ω a general polygon. In this situation $u \in H^{s+2}(\Omega)$, provided $f \in H^s(\Omega)$ where $s < \min(-1 + \frac{\pi}{\omega_N}, -1 + \frac{\pi}{2\omega_M})$ and ω_N and ω_M are, respectively, the largest interior angle at corners with non-mixed or mixed boundary conditions. For the special case of an L-shaped region with a non-mixed re-entrant corner $s = -\frac{1}{3} - \varepsilon$, so that $u \in H^{5/3-\varepsilon}(\Omega)$. For a region with a non-mixed slit, $s = -\frac{1}{2} - \varepsilon$ and $u \in H^{3/2-\varepsilon}(\Omega)$ so that from (2.10) at the slit vertex there is only one singular function in order that $u_\mu \in H^{2-\varepsilon}(\Omega)$

$$u_{j1} = r_j^{\frac{1}{2}} \chi_j(r_j) \sin \tfrac{1}{2}\theta_j \tag{2.15}$$

We remark that it follows from (2.13) and theorem 2.1 that for values of ℓ for which $\lambda_{j\ell} < \mu+1$

$$|c_{j\ell}| \leq \|u\|_{Z_{\mu+2}}$$

and hence from (2.14) that

$$|c_{j\ell}| \leq C\left\{\|f\|_{H^\mu(\Omega)} + \Sigma\|h_j\|_{H^{\mu+\frac{1}{2}}(\Gamma_j)}\right\},$$

showing that the $c_{j\ell}$ (stress intensity factors) depend continuously on the given data.

3. FINITE ELEMENT METHODS

In the previous chapter a Poisson problem in a polygonal domain with general mixed boundary conditions, together with its weak formulation in appropriate function spaces, has been considered, and the forms of singularities which can arise in the solution have been discussed. We now turn to the task of approximating solutions of these types of problems using finite element methods. Although we have thought it necessary to explain the function space setting for the general mixed boundary conditions, the finite element methods for treating singularities can, and thus will, be illustrated for the case of homogeneous Dirichlet boundary conditions. Problem (2.1) is thus simplified to become

$$\begin{aligned} -\Delta[u(\underline{x})] &= f(\underline{x}) , \qquad \underline{x} \in \Omega \\ u(\underline{x}) &= 0 \qquad , \underline{x} \in \Gamma_j, \ j \in \mathcal{D}, \ \partial\Omega_1 = \partial\Omega , \end{aligned} \tag{3.1}$$

with corresponding weak formation

$$\text{find } u \in H_0^1(\Omega) \ : \ a(u,v) = F(v) \quad \forall\, v \in H_0^1(\Omega) . \tag{3.2}$$

The Galerkin technique is applied to (3.2) by approximating $u \in H_0^1(\Omega)$ with $u_h \in S^h$, where $S^h \subset H_0^1(\Omega)$ is a finite dimensional space, and u_h satisfies

$$a(u_h,v_h) = F(v_h) \quad \forall\, v_h \in S^h . \tag{3.3}$$

Since $a(u,v)$ is continuous over $H_0^1(\Omega)$ and H_0^1-elliptic, it is well known that

$$\|u-u_h\|_{H_0^1(\Omega)} \leq C\|u - v_h\|_{H_0^1(\Omega)} \qquad \forall\, v_h \in S^h , \tag{3.4}$$

and also that, if S^h consists of piecewise polynomial conforming functions on a quasi-uniform mesh with size h, then the right hand side of (3.4) can be bounded using interpolation theory results so that

$$\|u - v_h\|_{H_0^1(\Omega)} \leq K\, h^{\gamma} |u|_k , \tag{3.5}$$

where γ depends on the order of the polynomial and on k. Inequalities (3.4) and (3.5) thus provide a bound on the $H_0^1(\Omega)$ norm of the error $(u-u_h)$. In view of the regularity theorem 2.1 and using the Poincaré inequality, if $u \in H^k(\Omega)$ then $|u|_k \leq C\|f\|_{H^{k-2}(\Omega)}$, so that the $H_0^1(\Omega)$ error norm in (3.5) can be bounded in terms of f. Bounds of a similar form have been devised, see e.g. Natterer (1975) and Nitsche (1975), for $\|u - u_h\|_{L_\infty(\Omega)}$, for which as a general rule one would expect the order of convergence to be one higher than that in (3.5).

The most important factor in determining the order of convergence is the regularity of u, as this in turn determines k. The analysis of Section 2 indicates that, for $f \in L_2(\Omega)$, when Ω is a convex polygon, problem (3.2) has solution $u \in H^2(\Omega)$. Thus, using spaces of conforming elements involving first order polynomials, one would expect $O(h^2)$ convergence for $\|u - u_h\|_{L_\infty(\Omega)}$. As has been shown for boundary shapes other than convex polygons, the level of regularity of the solution of (3.2) is lower (e.g. $u \in H^{5/3-\varepsilon}(\Omega)$, $u \in H^{3/2-\varepsilon}(\Omega)$), with the result that the order of convergence is reduced. Babuska and Rosenzweig (1972) using spaces of functions weighted with terms of the form ρ^γ, where different weights are used for the trial and test function spaces, have restricted the above low rate of convergence in the $H^1(\Omega)$- $(L_2(\Omega))$-norm to neighbourhoods local to the vertices of $\partial\Omega$.

For problem (3.1) in a polygonal domain with largest interior angle ω_N, Schatz and Wahlbin (1978) have derived inequalities of the form

$$\|u-u_h\|_{L_\infty(\Omega)} \leq K\, h^{\min(p+1,\ \pi/\omega_N)-\varepsilon}, \tag{3.6}$$

showing that the $O(h^2)$ convergence for problems with smooth solutions has been degraded by the corner (singularity); for problem (3.1) a corner with angle $\omega_N = 3\pi/2$ thus produces $O(h^{2/3-\varepsilon})$ convergence.

The above type of error analysis demonstrates the lower rate of convergence induced by the presence of a corner singularity. It also confirms theoretically the effect long known from numerical experiments, namely that a *standard* finite element method loses accuracy near such a singularity. This latter effect has prompted many workers to produce variants of standard finite element methods better able to treat the singular behaviour. It has already been indicated that the coefficient of the leading singular term(s) in the solution u, for example as in (2.8), is often a quantity with physical significance. Thus, whilst increasing the quality of the approximation, a finite element method for treating singularities should, if possible, also provide an approximation to this coefficient together with an error estimate. Several finite element variants for treating singularities are now discussed.

4. TECHNIQUES FOR SINGULARITIES

Comprehensive surveys of methods and elements for treating boundary singularities in two- and three-space dimensions, many motivated by computational techniques for linear elastic fracture, are given by Atluri (1980), Whiteman and Akin (1979) and Whiteman (1982), (1984). We do not seek here to provide a full list of methods, but rather to indicate techniques in the context of the three approaches for treating singularities: special elements, space augmentation with singular functions, local mesh refinement.

When the finite element method is applied to a problem of the type (3.2) using the Galerkin technique, if the region Ω is partitioned into elements e, equation (3.3) can be written as

$$\sum_e a(u_h,v_h)\Big|_e = \sum_e F(v_h)\Big|_e . \tag{4.1}$$

The usual procedure involves the mapping of each element in the physical (x,y)-space onto a *standard* element in (ξ,η)-space using a transformation of the form

$$\underline{x} = \underline{x}(\xi,\eta) = \sum_{i=1}^{q} N_i(\xi,\eta)\underline{x}_i \tag{4.2}$$

where $\underline{x}^T \equiv \{x,y\}$, N_i are basis functions local to the element and $\underline{x}_i$ are point evaluations of $\underline{x}^T$. For an isoparametric method the mapped image $\hat{u}(\xi,\eta)$ in the standard element of $u(x,y)|_e$ is approximated by

$$\hat{u}_h(\xi,\eta)\Big|_e = \sum_{i=1}^{q} N_i(\xi,\eta)(u_h)_i , \tag{4.3}$$

where the $(u_h)_i$ are point evaluations of the (unknown) trial function $u_h(\underline{x})$ at nodal points of the element. The bilinear form a(.,.) involves global derivatives u_x, u_y of u in physical space, so that under transformation (4.2) these are mapped into local derivatives $\hat{u}_\xi$, $\hat{u}_\eta$ so that

$$\{\partial u_{glob}\}_e \equiv \begin{bmatrix} u_x \\ u_y \end{bmatrix} = J_e^{-1}\begin{bmatrix} \hat{u}_\xi \\ \hat{u}_\eta \end{bmatrix} \equiv J_e^{-1}\{\partial u_{loc}\}_e , \tag{4.4}$$

where J_e is the Jacobian of the transformation. For each element the approximation (4.3) leads to global approximations u_h, where

$$\{\partial(u_h)_{glob}\}_e = J_e^{-1}\{\partial(\hat{u}_h)_{loc}\}_e . \tag{4.5}$$

Elements of most interest here are those which involve a point of singularity; it is usual for this to be at a node. For such elements the function $u_h(\underline{x})\big|_e$ is normally unable to provide an acceptable approximation to $u(\underline{x})\big|_e$; i.e. it cannot suitably reflect the singular behaviour. Thus some special action has to be taken to produce a suitable approximation for the singular case. From (4.5) it is clear that this can be effected through the Jacobian, see e.g. Henshell and Shaw (1975), Barsoum (1976), or through the local approximating function $\hat{u}_h$. For the latter, Blackburn (1973) and Akin (1976) proposed special singular elements, whilst Fix (1969) and Barnhill and Whiteman (1975) augmented the approximating space to produce suitable functions. Yet another approach is to refine the mesh near the singularity thus making u_h better able to model the singularity.

Perhaps the most used of the above techniques is one in which the singularity is introduced via the Jacobian. A frequently met form of singularity for problem (3.1) is that in which at the vertex of a slit, $\omega_j = 2\pi$, the singular functions have the form, see (2.9) - (2.11), $r_j^{\ell/2}\chi_j(r_j)\sin\theta_j/2$. In elements involving the vertex point the approximating function should have, in all directions within the element radiating from the point, the $r_j^{\frac{1}{2}}$-form of the dominant singular term. This can be achieved in certain elements with quadratic approximating functions by moving the nodes which are normally at the mid-points of the sides meeting at the singularity to positions on the sides one quarter of the lengths of the sides from the singular point. The effect of this node-shift is evident from the following *one dimensional* illustration.

We consider the element $[x_i, x_{i+2}]$ in physical x-space, with nodes x_i, x_{i+1}, x_{i+2}, as in Fig. 2, where $x_{i+1} = x_i + qh$, $x_{i+2} = x_i + 2h$, $0 < q < 2$,

Fig. 2

and the standard element [-1,1], with nodes at -1, 0, 1 in ξ-space. Transformation (4.2) for this case has the form

$$x(\xi) = x_i \frac{\xi(\xi-1)}{2} + x_{i+1}(1-\xi)^2 + x_{i+2}\frac{\xi(\xi+1)}{2} . \qquad (4.6)$$

Similarly the quadratic approximation (4.3) has the form

$$\hat{u}_h(\xi)\Big|_e = (u_h)_{i+1} + \tfrac{1}{2}\{(u_h)_{i+2} - (u_h)_i\}\xi + \tfrac{1}{2}\{(u_h)_{i+2} - 2(u_h)_{i+1} + (u_h)_i\}\xi^2. \tag{4.7}$$

The point x_{i+1} is one quarter the distance from x_i to x_{i+2} when $q = \tfrac{1}{2}$, in which case from (4.6)

$$\xi = \frac{2(x-x_i)^{\frac{1}{2}}}{h} - 1 .$$

Thus from (4.7) $u_h(x)\big|_e \equiv \hat{u}_h(\xi)\big|_e$ has the form

$$u_h(x)\Big|_e \approx A + B(x-x_i)^{\frac{1}{2}} + C(x-x_i) ,$$

where A, B, C are constants involving the element nodal values of u_h. An $x^{\frac{1}{2}}$-type singularity has thus been induced in $u_h(x)$ at $x = x_i$.

The $\tfrac{1}{4}$-point effect generalises to two- (and higher-) dimensions, with the following proviso. Whilst the $r_j^{\frac{1}{2}}$ form is exhibited radially for the six node quadratic triangle, for quadrilateral $\tfrac{1}{4}$-point elements this form is present only along radial directions coinciding with the element edges involving the corner.

The triangular $\tfrac{1}{4}$-point element has proved to be effective in solving problems of type (3.1) involving slits with $r_j^{\frac{1}{2}}$ singularities. To the best of our knowledge no asymptotic error analysis exists for the element which can be used to demonstrate improvements in the convergence rate. The element is attractive due to its ease of use, because it only requires the shift in the position of the relevant nodes. At the same time this technique demands that approximation $c_{j\ell}^h$ to the coefficients $c_{j\ell}$ be obtained by post-processing.

Another singular element is that due to Stern (1979), which is able to model a local field involving the leading singular term from (2.12) together with u_μ. O'Leary (1981) has analysed the error for this element for certain problems, showing that, if the singular elements surrounding a singular point shrink as the mesh size decreases, then the asymptotic rate of convergence remains degraded. The computational advantage of this approach is that the global stiffness matrix remains simple. For "not-too-small" mesh sizes the rate of convergence is in fact improved. With the Stern element the leading coefficients of $c_{j\ell}$ are approximated as internal variables $c_{j\ell}^h$, but convergence of $c_{j\ell}^h$ to $c_{j\ell}$ cannot be guaranteed.

As has been stated earlier, the technique of augmenting the approximating function space with singular terms was first proposed by Fix (1969) and later extended by Barnhill and Whiteman (1975). In this technique for elements involving the singularity the polynomial approximating function is augmented with singular terms. For simplicity we consider the case of triangular elements with linear approximating functions $u_h|_e$ for elements involving a singularity at the jth corner of $\partial\Omega$. In this case u_h has the form

$$a + bx + cy + \sum_{\lambda_{j\ell}<1} c^h_{j\ell}\, u_{j\ell}\ , \qquad (4.8)$$

involving extra (system) unknowns $c^h_{j\ell}$. Extra equations are needed in the global stiffness system to produce a nonsingular augmented global stiffness matrix. In the Galerkin technique $u_h \in S^h$ is then replaced by

$$\bar{u}_h + \sum_{\lambda_{j\ell}<1} c^h_{j\ell}\, u_{j\ell}$$

where $\bar{u}_h \in S^h$. It follows immediately for problems of the type (3.2) that the bounds (3.4), (3.5) are replaced by

$$\| u - (\bar{u}_h + \Sigma c^h_{j\ell} u_{j\ell}) \|_{H^1(\Omega)} \leqq K\, h\, |u_2|_2\ . \qquad (4.9)$$

It can be shown using the Céa Lemma and properties of the finite element spaces, see e.g. Ciarlet (1978) and Babuska and Aziz I1972) that

$$\sum_{\lambda_{j\ell}<\mu-1} |c_{j\ell} - c^h_{j\ell}| \leqq C\, h^{2-\mu-\varepsilon} \|u\|_{Z_2}.$$

Using theorem 2.1 the norm $\|u\|_{Z_2}$ can be replaced by $\|f\|_{L_2(\Omega)}$.

If a mesh is refined locally near a corner according to a prescribed manner, it is possible using only piecewise polynomial test and trial functions to prevent the rate of convergence from deteriorating from that associated with regions having smooth boundaries. Schatz and Wahlbin (1979) proposed a scheme for local mesh refinement to achieve this end, whilst Babuska and Rheinboldt (1980) indicated self-adaptive methods for implementing such techniques automatically. An advantage of such methods is that they do not demand a priori knowledge of the form of the singularity in the solution of a problem. Again the approximations $c^h_{j\ell}$ have to be obtained through post-processing.

5. A COMPARISON WITH BOUNDARY ELEMENTS METHODS

A related alternative technique to the finite element method for treating problems of type (2.1) is the boundary element method. In the present context the main advantage of the BEM is that only a one dimensional problem (on $\partial\Omega$) has to be discretised, thus leading to smaller linear systems than for the FEM. A general disadvantage of the method is that it demands a fundamental solution of the partial differential problem. The BEM has been adapted to treat problems involving boundary singularities, the variants being one dimensional analogues of those used with FEM; e.g. special elements Stern (1983), Xanthis et al.(1981), augmentation method papers by Costabel, Hsiao, Stephan and Wendland (1981), (1979) and Schleicher (1983). Complete error analysis is available for the augmented BEM, using the Galerkin procedure, as it is for the augmented FEM.

ACKNOWLEDGEMENT

The work of K.T. Schleicher on this paper was undertaken during a visit to the Institute of Computational Mathematics at Brunel University, financed by the British Council. This support is gratefully acknowledged.

REFERENCES

Akin, J.E. (1976) The generation of elements with singularities. *Int. J. Num. Meth. Engrg.*, **10**, 1249-1259.

Atluri, S.N. (1980) Higher-order, special and singular finite elements, Chapter 4 of "Survey of Finite Element Methods", A.K. Noor and W. Pilkey (Eds.), American Society Mech. Engrg.

Babuska, I. and Aziz, A.K. (1972) Survey lectures on the mathematical foundation of the finite element method, in "The Mathematical Foundation of the Finite Element Method with Applications to Partial Differential Equations", A.K. Aziz (Ed.), Academic Press, New York, 3-359.

Babuska, I. and Rheinboldt, W. (1980) Reliable error estimation and mesh adaptation for the finite element method, in "Computational Methods in Nonlinear Mechanics", J.T. Oden (Ed.), North Holland, Amsterdam, 67-108.

Babuska, I.N. and Rosenzweig, M.B. (1972) A finite element scheme for domains with corners. *Numer. Math.*, **20**, 1-21.

Barnhill, R.E. and Whiteman, J.R. (1975) Error analysis of Galerkin methods for Dirichlet problems containing singularities. *J. Inst. Math. Applics.*, **15**, 121-125.

Barsoum, R.S. (1976) On the use of isoparametric finite elements in linear fracture mechanics. *Int. J. Num. Meth. Engrg.*, **10**, 25-37.

Blackburn, W.S. (1973) Calculation of stress intensity factors at crack tips using special finite elements, in "The Mathematics of Finite Elements and Applications", J.R. Whiteman (Ed.), Academic Press, London, 327-336.

Ciarlet, P.G. (1978) The Finite Element Method for Elliptic Problems, North Holland, Amsterdam.

Costabel, M. and Stephan, E. (1981) Boundary integral equations for mixed boundary value problems in polygonal domains and Galerkin approximation, to appear in Banach Center Publications, Warsaw. Fachbereich Mathematik, Technische Hochschule Darmstadt, Germany, Preprint 593.

Fix, G. (1969) Higher-order Rayleigh-Ritz approximation. *J. Math. Mech.*, **18**, 645-658.

Grisvard, P. (1982) Numerical Treatment of Elliptic Problems with Singularities, Lecture Notes, ICPAM Summer School, University of Nice.

Grisvard, P. (1976) Behaviour of the solutions of an elliptic boundary value problem in a polygonal or polyhedral domain, in "Numerical Solution of Partial Differential Equations III", B. Hubbard (Ed.), SYNSPADE 1975, Academic Press, New York, 207-274.

Grisvard, P. (1980) Boundary value problems in non-smooth domains, University of Maryland, Lecture Notes No. 19.

Henshell, R.D. and Shaw, K.G. (1975) Crack tip finite elements are unnecessary. *Int. J. Num. Meth. Engrg.*, **9**, 495-507.

Kondrat'ev, V.A. (1967) Boundary problems for elliptic equations in domains with conical or angular points. *Trans. Moscow Math. Soc.*, **16**, 227-313.

Natterer, F. (1975) Über die punkweise konvergenz finiter Elemente. *Numer. Math.*, **25**, 67-77.

Nitsche, J. (1975) L_∞ convergence of finite element approximation. Proc. 2nd conf. on Finite Elements, Rennes.

O'Leary, J.R. (1981) An error analysis for singular elements. University of Texas, TICOM Report 81-4.

Schatz, A. and Wahlbin, L. (1978/9) Maximum norm estimates in the finite element method on plane polygonal domains, Part I and II. *Math. Comp.*, **32**, 73-109 and **33**, 465-492.

Schleicher, K.-T. (1983) Die Randelementmethode für gemischte Randwertprobleme auf Eckengebieten unter Berücksichtigung von Singulärfunktionen, Diplom-Thesis, Fachbereich Mathematik, Technische Hochschule Darmstadt, Germany.

Stern, M. (1979) Families of consistent conforming elements with singular derivative fields. *Int. J. Num. Meth. Engrg.*, **14**, 409-421.

Stern, M. (1983) Boundary integral equations for bending of thin plates, in "Progress in Boundary Element Methods Vol. 2", C.A. Brebbia (Ed.), Pentech Press, 158-181.

Wendland, W.L., Stephan, E. and Hsiao, G.C. (1979) On the integral equation method for the plane mixed boundary value problem of the Laplacian. *Math. Meth. Appl. Sci.*, **1**, 265-321.

Whiteman, J.R. (1982) Finite elements for singularities in two- and three-dimensions, in "The Mathematics of Finite Elements and Applications IV", MAFELAP 1981, J.R. Whiteman (Ed.), Academic Press, London, 37-55.

Whiteman, J.R. Problems with singularities, Sections II .6.0 and II.6.1 of "Finite Element Handbook", H. Kardestuncer (Ed.), to appear.

Whiteman, J.R. and Akin, J.E. (1979) Finite elements, singularities and fracture, in "The Mathematics of Finite Elements and Applications III", MAFELAP 1978, J.R. Whiteman (Ed.), Academic Press, London, 35-54.

Xanthis, L.S., Bernal, M.J.M. and Atkinson, C. (1981) The treatment of singularities in the calculation of stress intensity factors using the boundary integral equation method. *Comp. Meth. Appl. Mech. Engrg.*, **26**, 285-304.

INDEX

Adaptive mesh refinement 38, 71, 82-83

Adjoint operator 101

Advection equation 115

Affine manifold 129

Alternating direction method 34, 35-36

Approximation
- Galerkin 17, 73, 92, 108
- best 6, 17, 31, 41, 96
- best L_2 49, 54, 93, 119
- optimal 6, 18, 91, 93, 99, 105, 114, 132

Artificial viscosity method 64, 66

Aubin-Nitsche Lemma 7-9, 18, 21, 96

Babuska-Brezzi condition, see Inf-sup condition

Barycentric coordinates 49, 140

Basis functions 17, 19, 73, 98, 114
- hierarchical 73
- rational 159, 165-166

Biharmonic problem 42, 47, 50, 56, 154

Bilinear form 3, 16, 28, 33, 43, 47, 102, 112, 127, 128, 139, 148
- symmetric 5, 16, 92, 96
- unsymmetric 62, 94

Boundary approximation 21
- by piecewise conics 158-160
- by piecewise hyperbolae 158
- by piecewise C^1 curve 161-162, 164
- polygonal 157-158

Boundary condition
- Dirichlet 16, 42, 52, 58, 94, 170
- essential 16, 41, 43, 92
- interpolated 23, 43, 52, 157
- natural 17, 47, 57
- Neumann 17, 105, 153

Boundary element methods 169, 180-181

Boundary, inflow 96

Boundary layer 112

Boundary singularity 23, 169-181

Bramble-Hilbert Lemma 11, 19-21, 108

Bubble function 73

Cauchy Schwartz inequality 4, 44, 60, 66, 80

Cea's Lemma 6, 17, 20, 180

Central differences 97

Characteristic Galerkin method 115-120

Characteristics 115-120

Checkerboard pattern 52, 142

Coercive 5, 17, 92, 94, 96
- see also Ellipticity

Compact support 9, 60

Conductivity 82

Cone condition 11-12

Conjugate gradient method 34
- preconditioned 35

Conservation law 115

Conservative operator 115

Consistency condition 41, 45, 132

Continuity equation 51

Continuous-time Galerkin method 30, 31

Contraction mapping 2

Convection-diffusion problem 38, 61, 93, 95-105, 111

Convective velocity 95

Convergence 3
- of nonconforming method 44, 54, 64
- order 48, 76, 91, 95, 176
- weak 3, 9, 55

Corner singularity 83, 169-181

Correction indicator 71, 77-83

Courant-Friedrichs-Lewy (CFL) number 115-117

Crank-Nicolson method 115

Curl operator 51, 123, 124

Curved boundary 19, 23, 157-167

Defect correction 112-113

Diffusion coefficient 28, 33, 95, 103

Diffusion-convection problem, see Convection-diffusion

Dirac delta function 10, 16, 43, 63

Discrete-time Galerkin method 32-33

Distributions 9-12

Divergence operator 28

Divided differences 109

Dual space 4, 28, 94, 143

Edge singularities 169-181

Effectivity index 79

Ellipticity 45, 144, 172
- H^1- 53, 131, 172, 175
- H- 5
- S^h- 47
- V- 129,149
- V_h- 136

Embedding, dense 7, 148

Energy 41
- estimate 27, 29, 79
- inner product 5
- norm 41, 73, 79, 92, 106
- norm, discrete 43
- space 5

Error bound 6
- L_2 7, 31
- symmetric 27, 31

Error estimate
- L_∞ 16, 22-24, 36, 67, 176
- L_2 18, 23, 31-33
- optimal 27, 31, 33, 37, 101
- pointwise 106
- quasi-optimal 27, 31
- a posteriori 38, 71, 79-81

Error estimator 71, 77-81

Error indicator 71

Error, interpolation 19

Euler equations 119

Exponentially fitted scheme 98

Extrapolation operator 33

Extremal problem 91

Finite differences 32, 34, 46, 91, 97, 102

Finite element method
- conforming 1, 6, 15-23, 73, 92, 123
- Galerkin 15, 23
- hierarchical 72-77
- hybrid 123
- mixed 51, 57, 123-154
- moving 38
- non-conforming 19, 41-69, 104, 118
- Petrov-Galerkin 62, 98-106, 113, 120

Finite element
- basis functions 17, 19, 73, 98, 114
- bilinear 19, 103, 107, 115, 142
- C^1 41
- cubic 33, 37
- curved 19, 23, 157-167
- isoparametric 19, 23, 157, 160-164, 177
- linear 19, 93, 97, 109, 159
- nodal 43
- nonconforming bilinear 51
- nonconforming linear 48
- nonconforming Morley 50
- nonconforming quadratic 52

nonconforming Wilson 48, 50, 60-61
piecewise constant 54, 118
quadratic 19, 52, 56, 73, 159, 161-162, 178
quadrilateral 80

Flux 36, 107-109, 117

Functional 16
linear 2, 4
quadratic 17, 41

Galerkin method 27, 62

Gauss points 23, 36, 52, 107

Gauss quadrature, one point 52

Gramm matrix 31

Green's function 15, 23, 106, 111

Green's Theorem 45, 47, 65, 125, 126, 139, 150

Gronwall's lemma 29

Heat equation 35

Heaviside function 3, 7, 9, 63

Hyperbolic problems 38, 93, 114-120

Hypercircle 110, 112

Ill-conditioned 76

Incompressible fluid 41, 95, 123-125, 150

Inf-sup condition 129-131, 144
discrete 133-146
weak 149

Inner product 3
energy 5

Integration, numerical 18, 23, 36, 52, 146

Interelement discontinuity 81

Interior estimates 18, 23

Isometry 4

Jacobian 135, 158, 160, 163-164, 177-179

L-shaped region 83-85, 174

L_2 projection 32, 37, 116, 145, 146

Laplace operator 16, 56, 93, 123-125, 134-137, 167

Lax Equivalence Theorem 45

Lax-Milgram Lemma 5, 94, 96, 129, 172
generalised 63, 99

Lax-Wendroff scheme 115

Leapfrog scheme 115

Linear elastic fracture 177

Linear elasticity 123, 169

Linear functional 1, 100, 110, 127
bounded 2, 4

Linear manifold 16

Linear space
complete 3
normed 2

Linear transformation 20, 158

Lipschitz continuous 11, 94

Macro-elements 55

Mapping
bounded 2
continuous 2
contraction 2, 6

Mass lumping 34

Maximum norm 36

Mesh refinement 46, 61, 72, 177, 180

Metric space 1

Mid-side continuous 48, 51

Moving mesh 38

Multi-index notation 10, 15, 58

Multigrid method 76

Navier-Stokes equations 123, 154

Neumann problem 153

Non-self-adjoint problem 61, 91, 93-105

Nonlinear Parabolic equations 33

Norm 2
Dirichlet 107
energy 5, 17
L_1 120
L_2 9, 29

mixed 117
Sobolev 18, 37
weighted L_2 111
weighted Sobolev 22

Normed linear space 1

Optimal mesh 72, 83

Optimal recovery 23, 106-113

Orthogonal decomposition 152

Orthogonal projection 144

P-refinement 72-77

Parabolic Galerkin method 30

Parabolic equation 27
nonlinear 33-35

Parabolic smoothing 27, 30, 37

Patch of elements 46

Patch test 42, 46-69
generalised 42, 49, 57-61

Peclet number 64, 67, 97, 105, 112

Penalty method 143-148

Petrov-Galerkin method 62, 98-106, 113-120

Plane stress 82

Poincaré inequality 61, 172, 176

Poisson problem 16, 42-47, 49, 52, 107, 157, 169, 175

Poisson's ratio 47

Positive definite 5, 41

Potential energy 72

Potential theory 169

Precision, high order 165-166
linear 158, 160

Predictor-corrector method 33, 114

Re-entrant corner 174

Reduced integration 143-148

Reference element 19

Regular triangulation 134, 137

Regularized problem 143

Residual 73, 77

Riemann solver 119

Riesz Representation Theorem 4, 9, 98

Riesz map 4

Riesz representor 4, 6, 9, 102

Self-adaptive method 180

Self-adjoint operator 15, 17

Self-adjoint problem 72, 91, 100, 110

Semi-norm 10, 58, 108

Shape function 19, 73

Shock modelling 117-120

Singular functions 172, 177-179

Singular perturbation problem 105

Singularity
boundary 23, 169-181
corner 83, 169-181
logarithmic 43

Sobolev Embedding Theorem 11, 43

Sobolev inequality 11

Space
Besov 10
boundary 10, 171
complete 3
conforming 53
dual 4, 8, 28, 57, 94, 143
energy 5
Hilbert 3, 125, 127
$\underset{\sim}{H}(\text{curl},\Omega)$ 125
intermediate 10
nonconforming 53
pivot 8
primal 4
quotient 12
Sobolev 3, 9-12, 28, 57, 92, 170-171
Sobolev fractional 10, 37, 170
Sobolev weighted 22, 171, 176
tensor product 35, 36, 65
test 6, 16, 62-67, 92, 98, 104
trial 16, 57, 62-67, 92, 98, 114

Spline
natural cubic 113
quadratic 118

Spurious oscillations 62

Stability 41, 55
condition 132
of time stepping 34, 115

Stiffness matrix 100

Stokes equations 50, 123, 125-127, 137-143, 150

Strain energy 91

Stream function 51, 124, 152

Streamline diffusion method 103

Streamline upwind method 65-66

Stress 23, 42, 81, 91

Stress intensity factor 169, 175

Subspace, finite dimensional 41

Superconvergence 23, 36-37, 91-95, 106-109

Symmetrizer 99

Taylor-Galerkin scheme 118

Tensor product 23
   operator 36

Tensor viscosity 66

Test functions 17, 19, 59, 62, 65, 115, 117, 176
   ideal 100
   upwind 103, 119, 125

Time discretization 27, 32-35, 114
   Euler method 114
   explicit 34, 115
   implicit 34
   θ-weighted 32-34

Trace operator 10, 12

Traction jump 81

Trial functions 19, 62, 117, 176

Triangle inequality 45

Truncation error 33

Unit CFL property 115

Upwind differencing 97, 115

Upwind test functions 38, 65, 103, 119

Variational
   form 42
   formulation, two field 123-154
   principle 17, 41, 91, 123, 124

Virtual work 123

Vorticity 124

Weak form 5, 16, 28, 62, 91, 171

Weight function 2, 22, 105

Well-posed problem 15